CONCERNING THE HABITS OF INSECTS

CONCERNING THE HABITS OF INSECTS

BY

F. BALFOUR-BROWNE,
M.A., F.R.S.E., F.Z.S., F.E.S.

LECTURER IN ZOOLOGY (ENTOMOLOGY)
IN THE UNIVERSITY OF
CAMBRIDGE

∾

CAMBRIDGE
AT THE UNIVERSITY PRESS
1925

CAMBRIDGE UNIVERSITY PRESS
Cambridge, New York, Melbourne, Madrid, Cape Town,
Singapore, São Paulo, Delhi, Tokyo, Mexico City

Cambridge University Press
The Edinburgh Building, Cambridge CB2 8RU, UK

Published in the United States of America by
Cambridge University Press, New York

www.cambridge.org
Information on this title: www.cambridge.org/9781107600232

© Cambridge University Press 1925

First published 1925
First paperback edition 2011

A catalogue record for this publication is available from the British Library

ISBN 978-1-107-60023-2 Paperback

PREFACE

THIS book is the outcome of a course of lectures "adapted to a juvenile auditory" delivered at the Royal Institution during the Christmas holidays 1924. Its object is not so much to describe the life histories of various insects as to explain how these life histories were worked out, in the hope that others may be encouraged to do similar work.

There are few young people who do not take a fleeting interest in butterflies or beetles, and I venture to think that some of those who thus start as collectors may be stimulated to explore further when they see how easily this can be done.

The study of Natural History has a peculiarly elevating effect upon character, and, in these days, when education consists more of cramming than of culture, it is well to encourage a pursuit which is healthy both to the body and to the mind.

F. B.-B.

May 1925

CONTENTS

ILLUSTRATIONS

PLATES

MAPS

TEXT-FIGURES

Lecture I

INSECT COLLECTING AND WHAT
IT MAY LEAD TO

INSECT collecting is with many people an incident, like measles. They catch the complaint from someone else and usually recover and, having done so, they can look upon that period of their lives, in which they shamelessly ran about with a butterfly-net, with a thankfulness that there have been no after-effects. The broken and mouldy remains of their collections may be rediscovered from time to time amongst the rubbish of the house, to remind them of their transient weakness.

But there are some of us who never recover from this complaint and who, in fact, are proud of it and, in our unregenerate condition, we glory in spreading it to others and in seeing them caught with the enthusiasm.

It may occur to some of those among you to ask why should the insect collector be regarded by others with amusement and why should he usually be represented pictorially by a decrepit old gentleman in a slouch hat and dark glasses and with a green net chasing a butterfly. I cannot say why, but it is a fact that, not so very long ago in a case in the Edinburgh courts in which a will was in dispute, the fact that the testator collected insects was admitted as evidence of weakness of intellect.

Nowadays, although the collector is still regarded as not quite normal, the importance of insects in connection with man's affairs has come to be recognised, so that there is a more reasonable spirit abroad and I therefore feel justified in doing my best in these lectures to encourage you to take up insect collecting as an amusing pastime and in trying to show you how easy it is to study the habits and work out the life histories of these interesting creatures.

It is very difficult to say what it is that causes anyone to start collecting insects. It may be that, in some cases, it is merely a desire to acquire and accumulate something for oneself, the same instinct which starts off the novice in stamp-collecting, and insects, like stamps, are easily obtained. Enjoyment of beauty and artistic arrangement, added to the acquisitory habit, results in the production of a neatly arranged collection.

The individual who never gets beyond this stage may be classed with Darwin's 'Naturalists without souls' and he seldom shows much energy or ingenuity in his collecting. He buys or exchanges to obtain his specimens or, if he himself collects at all, he mostly goes to places where other people have already discovered the rarities. He is frequently a crank; perhaps he never puts anything in his collection except what he has collected himself and, lacking initiative, he resorts to artifices to fill up the gaps. For instance, I have heard of a collector who arranged with someone in another part of the country that, as soon as a particular rarity appeared, it was to be imprisoned on the spot where it was discovered and a telegram was to be sent so that the collector could come down and collect it himself.

On one occasion two collectors came to see me, as I had discovered some very rare water-beetles in Norfolk. It happened that I had two fine specimens of one of these rarities alive in an aquarium in my study and I offered these to my visitors. With very serious faces they told me that they never took anything unless they collected it themselves and I therefore gave them minute instructions with regard to the place where I had taken my specimens and they departed. At tea-time they returned empty-handed, having failed to find a single specimen, and I again offered them the live ones, suggesting that they might replace them in their collections when, later, they managed to find the species for themselves and, after considerable hesitation, each took a specimen, placing it alive in a glass tube.

After tea we visited a pond close to my house where another rather uncommon water-beetle occurred and, to my surprise, each of these collectors solemnly took from his pocket the glass tube containing the live beetle and, placing his net in the water, shook the beetle into the net and then lifted the latter and again placed the beetle in the glass tube. By this simple ceremony each of these gentlemen had now 'collected' this beetle for himself—from a pond where it had never occurred naturally and was never likely to occur!

Another fad of this type of collector is that he only takes so many specimens of each species, the explanation being that he has planned out his cabinet and there isn't room for more. This is always a mistake, if the collector hopes at any time to develop his collecting into something better because, in the study of any particular kind of insect, it is necessary to have plenty of material and from many different localities.

Now it may be gathered from the way I have criticized this class of collector—'the naturalist without a soul'—that he and his collection are useless and yet, if he will but take the trouble to label each of his specimens with the date of capture and the locality where it was taken, the material may prove very valuable to others, even if the collector never develops a deeper interest in it himself. I therefore, first of all, wish you to understand that I am not discouraging the mere collector but, rather, wishing him to continue his hobby, so long as he will accurately label his specimens.

Those of us who have the time usually discover that, besides reminding us of the glorious days in the country, the insects themselves lure us on to one or more lines of study. In the first place we usually start with attempting to name our speci-

mens and it is extraordinary how easy this usually
is to begin with and how difficult it becomes as
we accumulate knowledge. The naming of insects
is usually called 'systematic' work and most
collectors, however shallow their interests, usually
make some attempt in this direction. There are,
however, collectors who never acquire the capability of naming their specimens and who perpetually rely upon their friends or upon others
who have come to be regarded as authorities on
the particular group. This reliance upon others
sometimes leads to curious results. There was an
old collector in the south of Scotland who, when I
first knew him, was a collector of beetles and it
was he who first initiated me into the art of collecting. As a finder of beetles he had few equals but
he made little attempt to name his specimens,
which he used to send to a very well known
authority, the late Dr Sharp, one of the greatest
of British entomologists. The specimens were sent
in boxes, arranged in rows, and Dr Sharp used to
write out a list of names in the order of the specimens in the boxes. When the boxes were returned
to Scotland, the old collector would take out a
specimen and examine it lovingly and would perhaps replace it in the box in a new position so that
what was specimen 3 in row 1 might become
specimen 16 in row 7. In this way, as you can
see, what had been specimen 4 in row 1 now
became specimen 3 and received the name which
belonged to the specimen now removed lower

down, and thus this old collector used to record, from the Solway district of Scotland, species which were otherwise only known in southern England.

As I have said, naming insects is not so easy as it at first appears to be. I suppose that the novice first notices colour and shape and, not having studied even these two characters very carefully, two individuals of the same general colour and practically the same shape are at once regarded as belonging to the same species. This attitude is strengthened by the discovery that there are frequently slight differences in colour and shape between individuals of the same species, perhaps between the two sexes, but one does not take long to learn that both these characters are unreliable. This leads to a more careful study of form and it is really only by long and arduous searching that minute but reliable differences can be discovered upon which to allot to our specimens their correct names.

I do not propose to enlarge upon this part of the subject, as it is not capable of being properly discussed in a few words and, at least to those who are not already keenly interested in collecting, it is the dullest and most laborious part of the subject, while the other two lines of study to which collecting may readily lead are of much wider interest and will occupy us for the rest of the course.

From the very beginning of one's collecting experience, one notices that there is something peculiar about the way animals and plants occur. Most of them are not found scattered everywhere but are limited in their occurrence to certain kinds of surroundings such as woodland or open country, marshland or dry ground, lowland or high land and so on, and we find that those kinds which occur in one part of the country are, in many cases, not found in other parts. Broadly speaking, we can distinguish (*a*) limitation to a particular kind of surroundings or environment, and (*b*) limitation to a particular part of the country, and the observing and recording of the way in which the kinds of insects are distributed is a very interesting line of work which will justify the encouragement of insect collecting and which leads, as we shall see, to the study of much wider and more important problems.

Although there are collectors who try to collect all kinds of insects, most of us seriously study only those belonging to a particular Order, such as Butterflies and Moths or Beetles or Flies or Bugs and so on, and many of us find that the study of one of these Orders, even in such a small country as the British Islands, is more than it is possible to undertake. I have always been mainly interested in the beetles but, as there are more than three thousand different kinds in our islands, I found, after collecting for some time, that it was too large an Order to deal with and, having obtained a general idea of the groups within the Order, I at last settled down to a study of the water-beetles.

Now 'water-beetles' is rather a vague term, as representatives of various groups of beetles have taken to an aquatic life, but there are two groups which are most usually meant when water-beetles are referred to in the British Islands; these two groups together are represented by less than three hundred different species so that, by confining myself to a study of these, I have been able to obtain an abundance of information as to their distribution and habits, which I could not otherwise have done.

It is my object in this lecture to tell you something of what I have been able to learn by collecting water-beetles in the British Islands, so that you may realize how easy it is to get something out of collecting, besides, or perhaps rather than, a beautifully arranged collection.

In the first place one naturally seeks for water-beetles in water and, if one goes out into the country, one can find water in ponds and ditches, in lakes and in rivers and streams.

If we collect in ponds and ditches and compare the beetles we get in such places with those we get in lakes or in rivers, we discover at once that, although many of the same beetles may occur in all three, yet each has certain characteristic kinds, that is, we can definitely associate a number of water-beetles with ponds and ditches, others with lakes and others, again, with rivers and streams so that, at the outset, we find something interesting.

But how have we discovered this? Not by examining one pond and comparing our catch with that from one lake or one river, but by making great numbers of collections in different ponds, lakes and rivers and then comparing all the results.

We can speak of the 'community' of water-beetles occurring in ponds differing from the 'community' occurring in lakes and from that in rivers. But everyone knows that ponds are not all exactly alike, and neither are lakes nor rivers. There are small and large ponds, clear and muddy ponds, weedy and weedless ponds and so on, and there are lakes with gravelly beds and others with earthy beds; in some the water is clear and sparkling while in others it is brown and peaty. Similarly rivers may be large or small, swift or slow, clear or muddy and so on.

If we examine a number of ponds, keeping careful records of the beetles found in each, we shall very quickly find that the pond community of water-beetles differs in different ponds and we may roughly classify ponds into four types. (1) Ponds in lime or in gravel soil where the bottom is composed of very fine ooze or silt; these we can call 'silt ponds.' (2) Ponds in which the bottom is dark, thick mud composed largely of the remains of decayed vegetation; these we can call 'detritus ponds.' (3) Ponds in peaty soil where the water is usually tinted brown and is slightly acid; these we can call 'peat ponds.' (4) Ponds near the sea in which the water is usually slightly brackish or at least contains more salt than other ponds; these we can call 'brackish ponds.'

Each of these types of ponds really has a number of characteristics. For instance, the bog moss or sphagnum is associated with peat ponds, whereas various typical weeds, such as duck-weed and water-starwort and others, grow in detritus ponds. Silt ponds are usually characterized by comparative freedom from weed, though water-grass (*Glyceria*) is often to be seen spreading its blades upon the surface of the water.

Of the four types of ponds, the silt pond is by far the richest in the number of water-beetles which it contains, many of the kinds being only found in this type of pond.

But everything in this world is constantly changing and it is easy to observe the change of a silt pond into a detritus pond—but let us follow the history of such a pond from its birth to its death.

As an example of a silt pond I will take a marl-pit, an excavation for the purpose of extracting 'marl,' a mixture of chalk and clay, from which cement and bricks are made. There are many of these pits near Cambridge and as the land is low-lying, these excavations readily fill with water unless they are kept free by constant pumping. In such a pit, while still being worked, there are usually a number of pools in the hollows. These pools are at first bare of vegetation though, after a time, some begins to appear and as many as forty different kinds of water-beetles may be found in them.

Many of these excavations for marl extend over an acre or more of land before the work is given up and the pits may be forty or more feet in depth. When work ceases, the pumps are removed and the water gradually rises until it may fill the pit and turn it into a deep lake. If however the marl is only in pockets in the soil, small and shallow pits are dug out and these ultimately become filled with water. Vegetation gradually invades these pits, large and small, from the edges and as it dies off it produces vegetable detritus which gradually covers the bottom of the pond and in time fills it up, the length of time depending upon the depth of the water. Thus the silt pond slowly becomes a detritus pond and in time the latter becomes a marsh which finally dries up, and thus we see that, in speaking of a community of water-beetles characteristic of a particular type of pond, great numbers of ponds have to be examined and the results carefully compared, since no two ponds are likely to be in exactly the same stage in the process of change, so that no two ponds are likely to contain exactly the same community of water-beetles.

Just as all ponds are slowly changing in character and tending to become dry land, so are lakes, but the process is naturally very much slower and varies with different lakes. The reedy and boggy areas at the edges of lakes are evidence of this tendency and they provide pond conditions in which pond-beetles abound and from which the typical lake species are absent.

The detritus pond, although not containing quite as many species as the silt pond, is rich compared to the peat pool or the brackish pond, though each of these kinds contains its characteristic species.

The lake community consists of a few characteristic beetles which probably require a more equable temperature than pond species and perhaps also, for some reason, prefer water which contains more dissolved air and with a freer circulation, and in this connection it is to be noted that the river and stream community is very similar to the lake community, many of the water-beetles found in lakes being also found in rivers.

Now what started me off upon communities of water-beetles? It was just the fact that collecting them had led me to observe certain peculiarities of distribution which I investigated by keeping careful records of all the kinds of water-beetles I got in every piece of water in which I collected and, having accumulated great numbers of these records, I compared them with the records I had also made of the nature of the pond, lake or stream and found that there was a connection between type of community and the surrounding conditions. With the experience I have now had, it is possible to guess fairly accurately what kinds of beetles will be found in a particular spot by merely looking at it and noting the vegetation or the nature of the bottom and, on the other hand, it is possible to describe fairly accurately the type of pond, lake or stream from which a particular group of water-beetles has been taken.

I have spoken of the gradual change in the nature of the pond and the accompanying change in the community of beetles as if it were only a question of time, but it is sometimes possible to see the change by working a comparatively small area of ground and it is even possible to guess at the particular cause of the change in the community of beetles.

For instance, if one collects upon one of those flat areas so often found near the mouths of rivers where the sea-pink grows amongst the short turf and where there are usually many shallow ponds, areas which are liable to flood at spring-tides, one finds the pond community frequently dominated by a yellow mottled water-beetle known as *Agabus conspersus*. If the merse is such that part of it is never touched by the sea-water, *Agabus conspersus* is absent from the pools in that area and is replaced by a beetle somewhat similar in appearance known as *Agabus nebulosus*, a typical freshwater pond species. In this case then it appears that the amount of salt in the water is the determining factor as to whether *Agabus conspersus* or *Agabus nebulosus* is present and, as a fact, the former species is never found in our islands except in ponds round the sea-coast.

Here is another example where a different factor comes into play. In 1906 I was collecting one day on Goatfell in Arran and, climbing up from the Brodick side, I came to a peat bog at about 1000 feet above sea-level in which there were a number of peaty pools. I collected in many of these pools and I found that in all of them there were many specimens of a small dark-brown coloured beetle known as *Hydroporus gyllenhalii* which was the dominant species in the pools.

At about 1200 feet there is a similar peat bog with pools very similar to those lower down. Again I collected in many of these pools and found the same community of water-beetles as before, except that *Hydroporus gyllenhalii* was scarcely represented, its place being taken by a black species of about the same size, *Hydroporus morio*, which at the lower level had been entirely absent. Here, then, altitude was the apparent determining factor. We must not however assume that, having said this, we have solved the problem, because we must remember that every living thing is intimately related to its surroundings in that it has to feed, to reproduce and often to escape from enemies, so that the presence or absence of salt in the water or a slight difference in altitude may not directly affect the beetles concerned but may do so in some subtle way which still requires to be found out.

We can see therefore how collecting may lead us to a minute study of the relationship between the insect and its surroundings, and we may even be led to make experiments in order to test the conclusions arrived at by observations.

Before leaving this subject of water-beetle communities, there is one other point which one quickly discovers and that is that, just as in a community of human beings, different people do different things—for instance, while one is a grocer, another is a gardener and so on—so in a community of water-beetles, different kinds of beetles have different habits, so that all the members of a community are not actually struggling with one another to obtain the same food and the same place.

In a pond community, for instance, some of the beetles are to be found amongst the weeds while others are mainly bottom grubbers. The Squeak beetle (*Pelobius tardus*), for instance, although a member of the silt pond community, spends most of its time in the fine oozy mud at the bottom, its larva feeding upon a little red worm (*Tubifex*) which lives in burrows in the mud, while the Great Diving beetle (*Dytiscus* or *Dyticus*) never enters the mud except at the coldest time of the year when it does so in order to sleep.

Even in rivers and streams all the typical river species do not do exactly the same things. I remember, years ago, going to a particular burn in Dumfriesshire where a beetle known as *Hydroporus davisii* occurred. I had, at that time, never taken the insect but I had been told that it occurred in the 'well-burn' above Moffat. I spent a long time scraping up gravel from the bottom of the burn but with very little success and I had come to the conclusion that the little beetle had become very rare in this stream when, lying upon the bank and looking down into a clear pool, I saw numbers of individuals walking about upon the rocky side and, in a few minutes, I had as many specimens as I wanted.

On another occasion I went to the rocky sea-shore at Dalkey in Co. Dublin in order to find a little beetle, *Octhebius lejolisii*, which was said to be common there in the rock pools. Two of us spent nearly an hour there and only succeeded in catching one or two specimens and we were giving up, when I happened to look in a little pool hold-ing about a pint of water or less. There I saw several specimens and we then found that the beetle was abundant in these very small pools, too small to work the net in and, incidentally, this shows how one has to vary one's method of collecting in order to catch many of the kinds of insects. Observations on the habits of this little beetle showed that it has a very precarious exist-ence. It is only to be found in pools above high tide but it seems to like to be in a place where the salt spray can reach it. Now it is obvious that in a very sheltered area, the pools above high tide-mark which will occasionally get some sea-water will all be within a few feet of the high tide-mark whereas, on a stormy coast, pools up to a con-siderable height above the sea will get sprayed from time to time. Collecting on Clare Island off the west coast of Ireland, I found the beetle just a few feet above the sea on the eastern and sheltered side, whereas it was not until I got 100 feet above the sea on the western side that it occurred.

Because of the peculiarity of its requirements it has to put up with a good many changes in its surroundings. For instance, during a period of storm, the water in the pools will become pure sea-water while, during a period of calm weather, the water may become more salt owing to evaporation or it may become foul and stinking and may ulti-mately disappear altogether. Again, in such a calm period heavy rains will reduce the salinity and the water may even become fresh. The variations in the temperature of the water of these small pools is very great. On a hot day in August I have found a temperature of 86° F., while at night in that month the temperature sometimes falls almost to freezing-point.

One might well ask, why should any beetle remain under conditions which are so variable and apparently unpleasant, and the only answer avail-able is that the insect is suited to these surround-ings, not necessarily better suited to them than to nicer conditions, but perhaps excluded from the latter by other kinds. I made a few experiments with this little beetle and found that it did not live long in an ordinary freshwater aquarium and that the only conditions which it seemed to appreciate at all, other than its native haunts, were those obtainable in coke-breeze infested with bacteria, the material having been obtained from an unsatisfactory filter-bed in a sewage farm. One might expect, in the case of pond species, that water-beetles, being able to fly, the drying-up of the pond would cause a dispersal of the community but, from the rapidity with which a pond becomes re-stocked after water reappears in it, it seems that the beetles largely shelter in cracks in banks and in the mud when the water dries up. Proof of this is not easy to obtain because one cannot go about carrying supplies of water to fill up dry ponds during a drought, but indirect evidence is available. Within twenty-four hours of a pond becoming refilled with water, after a night's rain, I have found most of the community represented there and I have found this on many occasions. In one case, where a particular species occurred in only two or three of many pools on the top of a mountain, I found, after the pools had been dry for a fortnight, that the same pools contained the same species.

It may be asked, why not dig up the mud at the bottom of a dry pond and find the beetles in it? But this is much more difficult to carry out than it would seem to be. In the first place the beetles, when they descend into the mud, go to sleep, so that they do not readily move when the mud is disturbed. Secondly, they usually become coated in the material so that they become almost invisible, and thirdly, as I have said, under drought conditions such as cause ponds to dry up, refilling them artificially is a large problem.

I have so far discussed various observations which I have made in the course of a number of years' collecting and which indicate the possibilities of studying the relationships of insects amongst themselves and towards their surroundings generally, and now I want to discuss the subject of which the merest novice quickly learns something —that insects are scattered all over the country but that many of them are mainly or entirely confined to certain parts of it.

Whatever group of insects we start to collect we find that we can get different kinds by moving to different parts of the country, and by carefully recording the distribution of all the specimens in our collections we find that we can recognise four main groups of species. First of all there is a group of species which are found north, south, east and west in the British Islands and which we can call the 'general' type and we find that the majority of our insects belong to this group. There is no need to go to any particular part of the country to find them; work any district carefully and they will be found.

A small number of species are absent or very rare in the south but occur in the north and west, a group we can call the 'north and west' type, while another small group occurs only, or mainly, in the south and west, the 'south and west' type, and a still smaller group is found only in the south and east.

I have previously referred to the Great Diving

beetle (*Dytiscus* or *Dyticus*) of which we have six species in the British Islands, only two of these belonging to the 'general' type. One species, the 'Lapland Dytiscus' (*Dytiscus lapponicus*), belongs to the north and west type, having been found only in Scotland, mainly in the west, and in north-west Ireland. See map of distribution, p. 23. This is the smallest of our British species of this genus and is interesting, not only because of its limited distribution, but also because of the kind of environment in which it occurs. It is found only in small lakes, usually known as 'lochans' in Scotland, at an elevation of 800 feet and more above sea-level, and these lochans are characterized by having no stream running into or out of them and no trout are found there. There are usually newts and tadpoles in these lochans and often great numbers of the freshwater shrimp (*Gammarus*).

When one thinks about collecting water-beetles, one usually thinks of collecting by means of a net and at first one has the idea that all that is necessary is to work the net backwards and forwards in the water and then see what has been caught. Experience, however, shows that net-collecting itself is a fine art and that there are many different ways of handling and working a net which make all the difference between finding and not finding particular water-beetles. Experience also teaches us that water-beetles can be collected in many ways other than by net and we have already had one example of this in the case of the little

Octhebius. *Dytiscus lapponicus* is also much more easily caught without, than with, the water-net, the necessary experience being obtained by observing the habits of the insect. *Lapponicus* largely frequents lochans in which there are large stones in the shallow water and, so far as the daytime is concerned, it spends most of its time beneath these stones, coming to the surface at long intervals, twenty to thirty minutes or more, for a fresh supply of air and otherwise only swimming along the bottom from one large stone to another. If the lochan is weedy or has weedy corners, the net may catch specimens amongst the weeds, but most of the lochans where the species occurs afford little cover, except for the stones, and the net is practically useless. In such places there are usually spots where the water is not more than perhaps a foot in depth, and there the beetle can be got in quantities by quietly paddling about and very gently turning over the large stones and picking up the insects with the fingers as they swim slowly away seeking for a fresh shelter. Where it occurs it is usually abundant and one gentleman, to whom I had given instructions where to find it, wrote enthusiastically after visiting the lochan, telling me that he had taken eighty-nine specimens. I need scarcely say that he was one of those who never took more than eight or ten specimens for his collection! The species is, however, disappearing and has actually disappeared from one lochan where it used to occur. This lochan is situated

800 feet above sea-level in the top of a hill, a few miles behind Tobermory in the island of Mull. So far as I know, the last specimen was taken there in 1903 and I have visited the place two or three times since then, but without success.

Another example of this north and west group is a much smaller beetle known as *Deronectes griseo-striatus*. It is much more widely distributed than the *lapponicus* but is confined to Scotland and north and north-west Ireland. Other species of the type might be mentioned but these two are sufficient for our purpose.

The south and west group is represented by several species, usually small. One of these is the *Octhebius lejolisii*, already referred to, which has occurred all round the northern half of Ireland and will, no doubt, be found all round the south when it is sought for. It occurs on the south and west coast of Britain, as far north as the island of Skye where, however, it is by no means common. See map of distribution, p. 23.

Another member of this group is a pretty little reddish beetle with dark bands across it, known as *Bidessus minutissimus*, a species which has only occurred in a very few places in our islands, in the south of England, south-western Scotland, the Isle of Man and southern Ireland. This little beetle is very particular as to its habitat, being found only over a gravelly bottom in lakes or in shallows along the sides of rivers. See map of distribution, p. 23.

The south and east type is interesting, as all the

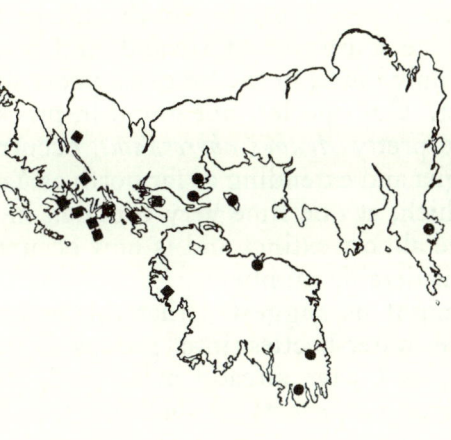

Map showing the distribution of *Ochthebius lejolisii* (black circles) an example of the 'south and west' type. It certainly occurs round southern Ireland but has never been sought for. Black squares represent the distribution of the Great Silver beetle (*Hydrophilus*), an example of the 'south and east' type. Its life history is described in Lecture V.

Map showing the distribution of *Dytiscus lapponicus*—an example of the 'north and west' type (black squares)—and of *Bidessus minutissimus*—an example of the 'south and west' type (black circles). The former is mentioned in the text and its life history is described in Lecture V.

members are confined to the south and east of Britain and the south-east of Ireland, and several of them are only found in the brackish pools along the sea-coast. One species, not found in brackish pools, is the pretty *Agabus abbreviatus*, occurring in East Anglia and extending as far north as York, a species which, at one time very common in the fens, became almost extinct and is now occurring again in considerable numbers.

Now it might be suggested that this arrangement of the water-beetles into groups is pure chance, but, as I have already mentioned, if we collect any group of insects we shall find that they will fall into the same groups and, in fact, the plants of our islands, the land and freshwater snails and the various groups of animals similarly are distributed into these groups, and this discovery is made by collecting and by keeping careful records of our results—and this is as far as collecting can take us, unless we go to the continent and continue our work there, when we shall find that our groups fit into groups which occur in Europe. For instance, our north and west group is composed of species which are found in northern Europe; our south and west group species occur in the Spanish peninsula and the western Mediterranean district, while our south and east group are mainly Mediterranean species and lastly our 'general' type can be described as mid-European.

We may conclude, therefore, that our grouping is not due to chance and we are naturally led to enquire how it has come about that our animals and plants are distributed as we find them.

The first attempt at explanation seems to have been made by Prof. Edward Forbes[1] in 1846. He pointed out that our animals and plants had obviously come to us mainly from the continent of Europe, of which our islands are only a small outlying portion, separated from the mainland by very shallow seas, deep water commencing a little to the west of Ireland. He further showed that we have scarcely any animals or plants in our islands which are not found on the continent and he explained our different groups by assuming that they had come to us at different times.

Now the groups or types to which I have referred are not quite those first recognised by Forbes but they are sufficient for our purpose and they have received names according to their supposed sources of origin judging by their present distribution in Europe.

The south and west group, called the 'Asturian' or 'Lusitanian' type, because the same species exist to-day in the Spanish peninsula, whence therefore it is supposed to have come to us, is represented by fewer kinds of animals and plants than the other types, and it has been regarded as the oldest type on the ground that it is the remnant of a fauna and flora which once covered the whole area

[1] 'The Geological Relations of the Existing Fauna and Flora of the British Isles, etc.' *Memoirs of the Geological Survey*, vol. i, 1846, pp. 336–432.

of these islands and has been squeezed out, first by the north and west group, the alpine or arctic or Scandinavian type, which in its turn has been squeezed out by the 'general' or Germanic type.

As to the time when these different types arrived, it has been argued that the arctic type, from the fact that it exists at the present time in the arctic regions and northern Europe, came to us from the north and must, therefore, have come at a time when the climate of our islands was colder than it is now. We know that Britain was, at one time, some millions of years ago, under a tropical climate when lions, hippopotami and other animals now living in the tropics, existed here, and we also know that, since that time, our islands have passed through an ice age when great glaciers moved down the mountains to the sea. The arctic type has therefore been regarded as having come to us with the ice age, which means that the Lusitanian type came to us before that period.

But, as I have already said, before the ice age our climate was tropical so that in the change from tropical to arctic the country would have become suitable for animals and plants preferring a temperate climate and, as fossil evidence shows us, when we were enduring a tropical heat the arctic regions were enjoying a climate not unlike that which we have now. Thus the cooling down of the arctic regions, which of course took place at the time when we were cooling down from tropical temperatures, would have tended to drive the temperate animals and plants southwards from the polar regions and our pre-glacial so-called Lusitanian type species must therefore have reached us from the north and not from the south.

As the climate became colder, these temperate forms were either killed out or were squeezed into corners where conditions were not impossible for them and the country, so far as it was habitable, was mainly occupied by the arctic invaders. With the cessation of the ice age and the gradual return of temperate conditions, these northern forms had to retreat, being driven north and west by new invaders coming in from the south and east and our main group of animals and plants, the Germanic type, is the post-glacial invasion, the south and east group being the most recent arrivals from southern Europe, which are likely to become a more numerous group, if we again approach tropical conditions.

In this lecture I have endeavoured to show how mere collecting of any one group of insects may lead one to examine problems which are bound to arise, if one keeps careful records of the work done, and I have really sketched for you my own biography and how I was led to study problems of insect communities and of geographical distribution. Now I do not expect everyone who collects insects to have the time or the inclination to study such problems, but mere collecting, so long as careful and methodical records are kept and made available for the use of others, is good and may ultimately assist in clearing up some of the complex problems of natural history.

In these studies, I have not relied only upon my own collecting but have used to the full the published records of innumerable naturalists who have, perhaps, never had time to do any more than spend glorious days in the country in the healthy enjoyment of catching insects—people who would frankly admit that insect collecting was their hobby.

And these records of amateur naturalists are not only of use in coming to conclusions as to the past; they are going to be of use in the future because, from them, we not only know what was, but later on we shall be able to observe the changes that have taken place, and thus obtain new evidence on the question of distribution.

I hope, therefore, that those of you who never do more than collect insects as a hobby will be encouraged by realizing that your results, however small, may really be a contribution to science.

Lecture II

THE HABITS OF BEES AND WASPS

IN the previous lecture I endeavoured to show how interesting collecting might become if we kept careful records of our work and speculated as to the reasons why insects were distributed in the way we find them, and now I want to discuss another line of work which usually arises out of collecting and often supersedes it in interest, and that is the study of life history. When one visits a particular spot where some rare insect is to be found, it may be that, at the time of the visit, the insect is unobtainable in the adult condition and, in such a case, if one can find the young stage, for instance, the caterpillar if it be a butterfly or moth, one collects a number of specimens, feeds them and thus ultimately obtains the adults—and this is frequently the beginning of interest in the study of life history. We watch the caterpillar changing its skin, we watch it spinning its silken threads, we notice that it changes into a pupa or chrysalis and that, after a longer or shorter time, the pupa breaks open and the perfect insect emerges, at first damp and soft, with crumpled wings, and we watch the expansion of the wings and wonder how long we shall have to leave the precious specimen before we may kill and set it.

Perhaps only two or three of the insects emerge from the pupal skin but we are fortunate enough to get a batch of eggs from them so that, by carefully preserving these through the winter, we shall be able next year to rear a large number of specimens. We are thus acquiring an interest in what the insect does and this may easily develop into a study of habits.

Now, studying habits in insects requires a number of virtues in the student. In the first place it entails patience—it may mean watching for hours at a time and discovering very little. Secondly, it requires great concentration of attention and the noting of the most trivial facts in the insect under observation and, in the third place, it requires considerable ingenuity in devising means of watching insects which, under ordinary circumstances, do not admit of watching, and again it requires judgment in drawing conclusions from what one sees.

In saying all this I have perhaps discouraged many of you from attempting this very interesting line of work but, in this lecture and those which follow, I hope to show you that it is not as difficult as it seems to be and that, to a large extent, this sort of work is not unlike something which we are mostly rather good at, and that is 'teasing.' Now teasing other people is something which is not encouraged by our elders but, although it is itself reprehensible, the principle underlying it is a good one. When we tease somebody, we are doing something with the object of seeing what that other person will do—which is in the nature of a scientific experiment since we often experiment upon living things with the object of seeing how they will react. So, when we undertake the study of life history and habit we tease or experiment with the insect in order to see what it will do—and perhaps this light upon this line of study will encourage some of you to undertake it.

In this lecture I shall tell you what I have learnt about certain bees and wasps and how I was able to study them, but first let me say a few words concerning bees and wasps. The words convey to your minds two or three distinct insects: the humble-bee and hive-bee and a black and yellow wasp which is usually a great nuisance in the autumn. But there are, in reality, great numbers of bees and great numbers of wasps, all of which, far back in the dim distance, originated from a common ancestor.

After these two groups became distinct, they developed along more or less parallel lines and thus we find, both in bees and wasps, what we call 'social' species in which a whole family lives together and the work of the home is divided up amongst the individuals. But these social forms in every case arose from other forms which, in order to distinguish them, we call 'solitary,' in which each individual attends to its own business and there is no communal life. Many of these so-called solitary forms live in colonies where hundreds of individuals may be found together, but each one does its own work, regardless of its neighbours, each female builds her own nest and resents interference by others.

We have established then, in a very general way, the difference between social and solitary forms of bees and wasps and, as evidence that the former have originated from the latter, it is possible even at the present time to recognise intermediate forms.

My interest has been mainly centred in the solitary forms, of which there are a considerable number in England, and although I had casually observed many of these from time to time, it was not until 1917 that I first began to study some of them seriously. Since my object in this lecture is to show you how such a study may be undertaken rather than to recount the habits of the various kinds I have worked with, I shall concentrate attention upon only a very few types and as a first example of a solitary bee I shall take a very common British species known as the Red Osmia or *Osmia rufa*, *v*. Plate 1.

This bee reminds one of a small humble-bee in its shape and hairiness and, except for its head which is black, it is reddish brown in colour, owing to the body being covered with a dense coating of reddish brown hairs. In May and June, this bee may be found visiting the flowers in the garden or crawling about upon the garden soil, especially in places which have been recently dug.

I first became interested in its doings when I

had a small garden in Cambridge and my curiosity was aroused by its habit of settling upon the damp soil, moving about for a few seconds and then, apparently, digging its head into the ground, after which proceeding it flew away over the fence. By watching this digging performance I discovered that the bee was gathering in its jaws a small pellet of earth and removing it, and I was then curious to know what it did with this. I therefore read all I could about the insect and I learnt that it used the earth to construct a nest which it built in holes in banks, nail-holes in walls, holes in trees, key-holes and all sorts of places.

As it appeared that there was no suitable place in my garden for it to build—except the lock of the summer-house door, which I found was occu-pied—I set about supplying the deficiency and I therefore constructed a shallow wooden box about 2 feet long by 18 inches broad and 3 inches deep such as gardeners use for seeds and cuttings. This I completely filled with wet clay mixed with chopped straw, so that the material should not dry too hard. In the surface of this I bored a number of holes about $\frac{3}{4}$ inch in diameter, these holes reaching to the bottom of the box, so that in this way a number of burrows 3 inches long were made. The box was then set in the sun to dry and it was then fixed vertically upon the garden fence facing the south, so as to get as much sunlight as possible, because all bees and wasps are sun-lovers and do little or no work in the absence of sunshine.

I had not long to wait before the Osmias and other bees and wasps came to prospect as to the suitability of the accommodation offered. In a few days a few individuals had begun to work in the burrows and I soon had the satisfaction of seeing them bringing in loads of pollen from the flowers, and pellets of earth, the burrows being ultimately closed by having the entrance filled in. This was satisfactory so far as it went, but I wanted to see more; I wanted to know what happened inside the burrows, and I wanted to see what was going on inside after the entrance had been closed. I could, of course, obtain the contents of a burrow by cutting into the dry clay, but this caused the clay to crack and crumble and seriously upset the work going on in other burrows, so I had to devise another scheme. I made a number of short wooden tubes by cutting pieces of elder and boring out the pith. The tubes I plugged at one end with a little ball of pith or paper and the tubes were then tied up in bundles and laid on a shelf fixed upon the garden fence. At the same time I cut a number of glass tubes 3 or 4 inches long, plugging them in the same way as the elder tubes. Each of these glass tubes was then placed in a tube of brown or black paper from which it could be easily withdrawn, the paper sheath being for the purpose of making the tube dark inside, *v*. Plate II. The glass tubes were also tied into bundles and laid on the shelf.

These tubes attracted the bees and a number

PLATE I

A 'bee wall' showing bricks, holding the elder and glass tubes. A number of elder tubes can also be seen on the left resting in a rack, and a few can be seen in the middle of the picture set perpendicularly.

Some of the bees which visit my 'bee wall': (1) *Osmia rufa*; (2) the leaf-cutter (*Megachile willoughbiella*); (3) *Anthophora personata*; (4) *Cœlioxys conoidea*, which lays its eggs in the cells of the leaf-cutter; (5) *Melecta armata*, which lays its eggs in the cells of Anthophora; (4) and (5) are called 'cuckoo-bees' because of this habit.

PLATE II

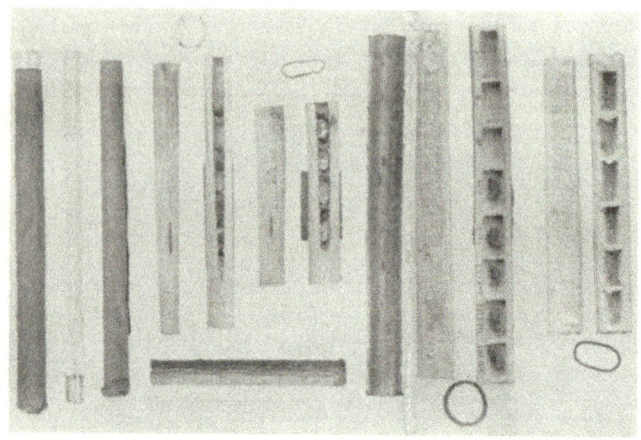

Glass tubes and elder tubes as used in the 'bee wall' and cells cut in elder pith to which the contents of the elder tubes are transferred when these tubes are split open. The rubber rings are for holding together the halves of the split tubes.

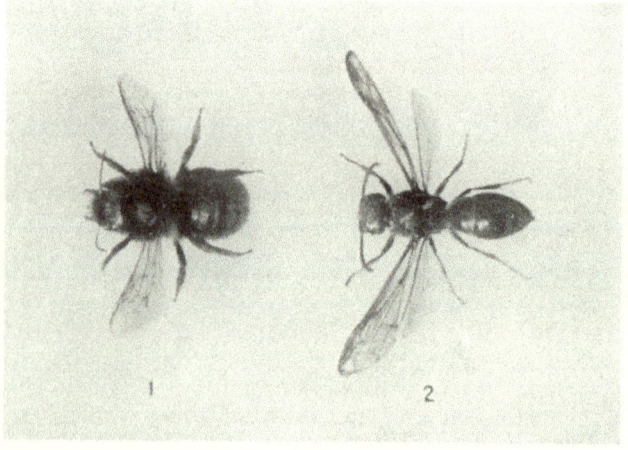

The Red Osmia (*Osmia rufa*) and one of the Odyneri (*Odynerus antilope*); the bee and the wasp mainly dealt with in Lecture II.

of them were occupied and closed, but the method was inconvenient because the removal of a closed tube from a bundle caused the displacement of the other tubes so that a considerable amount of confusion arose amongst the bees working there and disputes as to ownership occurred amongst them. I ultimately got over this difficulty by obtaining a number of 'ventilator' bricks, bricks pierced by three rows of seven, eight or nine holes. These were once used by builders as ventilators in walls beneath floors. Into each of the holes in the brick I put a glass or an elder tube and set the brick up on its side on the shelf and by this means I was able to remove any tube and replace it without disturbing the others, v. Plate 1. As soon as a tube was closed by a bee I removed it and replaced it by a new one, and thus there were always a large number of suitable residences available. I kept a record of everything I did, even the apparently trivial fact of the number of tubes I removed each day from these bricks, and this record has given me a rather interesting result as, by arranging it as a chart, it shows the period of activity of the Red Osmia. I give the chart for the year 1919 as being the most complete (v. Text-figure 1) and it will be seen that, leaving out of consideration for the moment the days on which the bees closed few or no tubes, the first tubes were closed on April 12th and that the bees became more and more active until May 22nd, when twenty-five tubes were closed, after which activity gradually diminished until the last tube was closed on June 14th and after that there were no Red Osmias to be seen about.

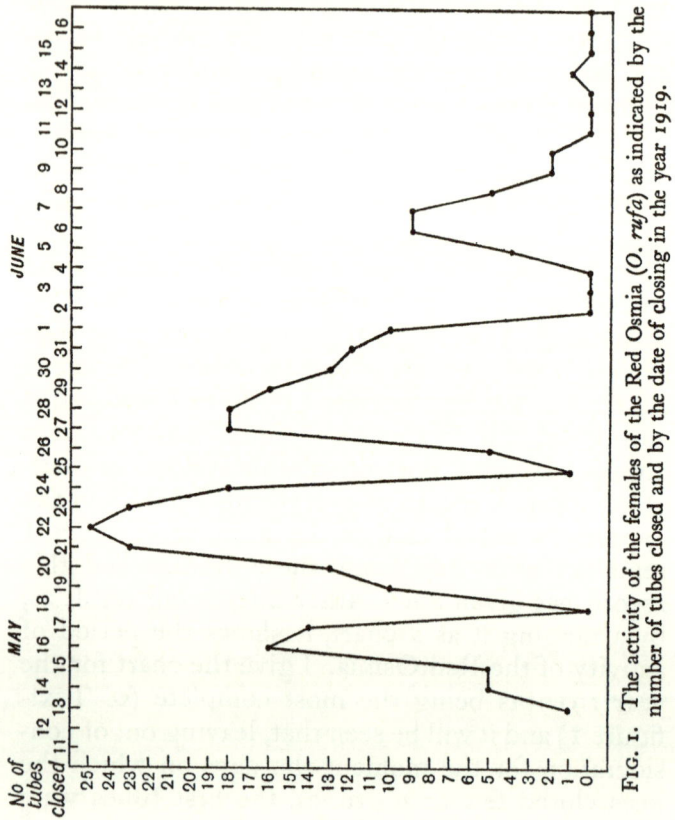

Fig. 1. The activity of the females of the Red Osmia (*O. rufa*) as indicated by the number of tubes closed and by the date of closing in the year 1919.

Now why did the bees do little or no work on certain days during their active period? Fortunately I had kept a record of the weather all through this period and from it I found that May 18th was cold and dull and the 19th was cold but bright. May 25th was, I believe, wet, while in the beginning of June, although there was bright sunshine, the wind was cold and in the north-east. We see, therefore, that sunshine alone is not enough to keep the bees at work, but the temperature must also be suitable.

Thanks to the extraordinary concentration of the Red Osmia upon its work and its almost complete disregard of anything else while it is at it, I was able to observe everything that happened in a glass tube, once it had been occupied, by simply withdrawing the tube from its paper sheath. The sudden flooding with daylight did not in the least disturb the working bee so that, as soon as she had returned home, I merely had to pull out the tube and watch what she did and I will describe what I saw.

The bee, having selected a glass tube to her liking and got to know its position, began work by bringing home a pellet of earth. She entered the tube head first and, crawling inward, she deposited the pellet upon the floor or side of the tube, working it with her jaws so as to spread it out across the wall of the tube. This was the commencement of the innermost wall to be built in the tube, a wall which is seldom placed against the plug at the inner end but usually a short distance from it. Having deposited the first pellet, a second one was brought and this was placed and worked alongside the first, more pellets being fetched until the wall completely closed in the end of the tube. So far, not much work had been put into the building, beyond the joining up of the different pellets, but now additional pellets were brought and carefully worked upon the face of the wall so that a concave and comparatively smooth surface was produced.

The number of pellets used in this wall varied roughly from ten to fifteen, and after it was completed the bee brought in one or perhaps two more, and she then definitely marked out the position at which the next wall was to be built by plastering this material on the wall of the tube about $\frac{1}{2}$ to $\frac{3}{4}$ inch in front of the other wall, thus marking the limits of the cell which will be completed later. But before it can be completed it must be stocked and the bee therefore now becomes a forager, visiting flowers and sipping the nectar which, in her crop, is quickly converted into honey. In pushing her way into the flowers to reach the nectar her body becomes smothered in pollen from the stamens and this adheres most readily to her under side where the beautifully branched hairs are specially long.

And now an interesting observation is made, as a result of carefully watching the return of the foraging bee. Different kinds of flowers have different coloured pollens which may be pale cream, yellow, orange, red or purple. The Red

Osmia always returns from a journey with only one coloured pollen, showing that she has only visited one kind of flower or, at most, several closely related kinds. If, however, the pollen stored in the completed cell be examined it will not infrequently be found to have streaks or patches of different colours in it, and continued observation of a particular bee shows that she may visit different flowers on different journeys.

The bee, on her return from a foraging expedition, enters the tube head first and, proceeding close up to the completed wall, she disgorges the honey from her crop. This only occupies a few seconds and she then turns round in the tube if she is not too large—which she generally is— in which case she backs out, turns round at the entrance and backs in and, by means of the brushes on her hind legs, she removes all the powdery pollen from her body. Having thus cleaned herself she sometimes turns round once more and mixes up the honey and pollen with her mandibles, but frequently she omits this and starts off upon another journey for further supplies.

When the cell is half-filled with the accumulated pollen paste, the bee ceases work for a time and then, disgorging a drop of honey upon the centre of the exposed face of the paste, she turns round and lays an egg upon it, firmly fixing one end of the elongated egg in the wet patch, *v.* Plate III.

After this, the bee once more becomes a builder and constructs the outer wall of the cell at the line already marked out. This wall, like the first one, is concave and smooth on its outer surface and serves as the inner wall of a second cell whose length will be marked out in the same way as was that of the first. Thus a series of cells—even as many as fifteen—may be built in a tube if it is long enough, though in my tubes, which are mostly 3 inches long, four cells are the usual number. The outermost cell of the series is almost always left empty and the final wall is usually built at the entrance to the tube, this wall being about twice the thickness of any of the others and constituting a 'front door.'

In about ten days the egg hatches but the shell is so thin that long before this the bee grub or larva can be seen developing within. When the hatching takes place, the grub does not move from the spot where the egg was laid, the thin shell merely splitting along the back of the grub and shrinking until it merely forms a kind of mat surrounding its base. For a short time the grub rests and then bending its head and body downwards it begins to feed upon the pollen paste, gradually eating it away and actually undermining its own position so that it ultimately falls to the floor of the cell after perhaps five or six days.

The grub grows rapidly, casting its skin four or five times during its life, the skin always shrinking underneath it, as did the egg-shell, and in from four to eight weeks the larva is full grown, having during that period exhausted the food supply in the cell. It then begins to spin a silken cocoon, the silk being spun at first loosely in the cell and later very compactly so that the larva is ultimately enclosed in a blunt-ended reddish-brown case, in which it rests for a number of weeks, after which, casting another skin, it changes into a white pupa. This takes place in July or August and for some weeks this apparently lifeless stage remains quite unchanged. Then the large eyes assume a reddish tinge and gradually the whole pupa darkens and becomes black, and at the end of September or during October the thin pupal skin splits and the adult bee emerges from it within the cocoon. There it remains, sleeping through the winter and only reawakening in the spring when it bites its way out of the cocoon into the cell.

In order to complete the account of the life history we must say a few words about the escape of the bee from the cell because, you will remember, there may be a series of as many as fifteen cells in a tube from which there is only one way of escape. If the bee in the outermost cell of the series is the first to free itself, it has merely to bite through the outer wall, pass through the empty cell and then break through the front door, but supposing the bee in the innermost cell is ready to escape before any of the others? The great French naturalist Fabre has shown in the case of another Osmia that an individual will die rather than break through the next cell, if that next cell contains a living bee. If, on the other hand, the occupant of that cell is dead, the imprisoned bee will bite her way through the cell, the cocoon and the carcase in order to escape. Thus, if the bees in the outer cells are alive, those in the inner cells wait patiently until the way is clear for them to pass out and, if the delay is too long, they quietly die where they are, perhaps trying first to gnaw through the side wall or pass round the intervening cocoons.

I have outlined the life history of the Red Osmia from the time the mother commences to build her cell to the time that the occupant of that cell escapes as an adult bee and this life cycle occupies almost a year, but all the observations I made upon the eggs and grubs were not made upon the comparatively few individuals which grew up in glass tubes. I have mentioned the elder tubes I supplied for the use of the bees and these were occupied much more readily than the glass tubes. By splitting these elder tubes, after the bees had closed them, it was possible to extract the contents of the cells and by cutting out a number of cells in the pith of split elder, as shown in the top figure of Plate II, I could place the pollen paste and the egg of the bee in these and observe what happened even more easily than I could see the doings of the occupants of the glass tubes, because the glass sometimes became dull on the inside and thus obscured the occupant.

You may wonder, however, how I found out the length of time the larva lay in its cocoon before changing to a pupa and how long the pupa rested before becoming a bee. The removal of a larva from its cocoon was always fatal to it and, with very few exceptions, a pupa also died under the same circumstances. I therefore cut a small window in each cocoon, the cocoons thus treated all being kept in small cardboard boxes with tightly-fitting lids. By opening the boxes every few days, the air supply was kept sufficiently good and by looking through the hole in the wall of the cocoon it was possible to see the occupant. The necessity for keeping the cocoons in closed boxes was due to the fact that the delicate pupa dries up very quickly if exposed to the air, and those which I kept on damp cotton-wool also died very quickly and became covered with mould.

The first individuals of the adult bees to make their appearance each season are the males, for the most part smaller than the females, with greyish-white hair in front of the face, the female having a black face. The male also lacks two projections from the front of the head which the female possesses. These males may be about early in March, sometimes weeks before any females appear, and they spend their time feeding and hovering round the burrows from which they expect the females to emerge. At night they sleep in the tubes and, at intervals during the day, and on cold and dull days, they rest, head to entrance, in them. The whole of the life of the male is spent in feeding and courting—in fact the males of most of the bees and wasps are rather degenerate individuals and the time may come when they will cease to exist and the females will manage without them.

The stress of life falls upon what we human beings have been wont to speak of as the 'weaker sex' and, although, no doubt, the female takes a certain amount of relaxation, most of her time is occupied in providing homes and food supplies for her offspring which she is not destined to see. Even from the moment of her appearance from her natal tube she takes things seriously in that on her first flight she makes a careful inspection of the place from which she has emerged, possibly considering, even then, the prospects of later on planting out some of her own family, for this bee seldom makes new galleries for herself, preferring to use what she can find, whether they be ones used previously by her own species or by some other; in fact any suitable tunnel, whatever its origin, is made use of.

I don't know whether, under natural conditions, the Red Osmia is very particular in making her choice but, in the case of my tubes, the indecision of individuals is most trying to one's patience. A bee will go from tube to tube and back on the same ground over and over again before finally selecting a home and, even then, it is by no means certain that she will not change her mind.

Once the choice is definitely made the bee backs in and rests for some time in the chosen tube and, very frequently, I have seen her roll about in the tube, as one may see a dog roll on the grass.

I could usually be certain, after seeing this performance, that she had chosen the site for a series of cells, and I imagine that the object was to convey to the tube her own body-scent so that, on her return, she might recognise the place. Her next actions were with the object of learning the situation of the chosen tube by flying backwards and forwards in front of it and facing it, quite obviously noting landmarks by which she might find it again. Such a flight might occupy fifteen or thirty seconds and would begin by a minute inspection of the entrance itself, the flight gradually extending farther on either side, after which the bee would either re-enter the tube or disappear for a short time and then return, repeating the forward and backward flights in front of the entrance. After such a lesson she seemed to remember the position though, as in human beings, it was easy to recognise that some individuals learnt more quickly and accurately than others and sudden loss of memory on the part of an individual was not uncommon, even after a tube had been worked at for some time. Consequently frequent scuffles were to be seen, an indignant owner trying, usually apparently with success, to oust an individual which had entered the wrong house, but I never saw any injury caused through these disagreements, although both bees would frequently fall to the ground locked together, when they would at once separate and return as quickly as possible to the scene of the dispute, the first to arrive backing in hastily with the object of holding the castle against the other.

In order to observe the habits of individuals it was necessary to be able to recognise them, and although most probably each bee has a distinct appearance—some were easily recognisable—my acquaintance with them was not so intimate as to enable me to recognise them all. I therefore marked them on the back with small spots of different coloured oil paints and for this purpose each bee had to be caught, which was an easy matter, by taking it in the fingers as it arrived to enter a tube or emerged from one. The Red Osmia does not sting, even when being handled, and I had no difficulty in holding an individual between finger and thumb and applying a fine brush bearing the oil paint.

At first I was rather clumsy in this marking, and occasionally the paint not only touched the back but got upon the wings so that, when released, the bee fell to the ground, unable to fly. However, it seemed very little disturbed and, crawling about until it found a support, it climbed up this and cleaned its wings, using the brushes of its hind legs. Having done this, it flew away as if nothing untoward had happened, returning again shortly to continue its work.

If I caught a bee as it returned laden with pollen, its first act, upon being released, was to fly to some near-by support and carefully clean itself of all the pollen which had been contaminated by the touch of my fingers, after which it would fly away and gather a fresh load. None of the bees showed any fear, nor did they get unduly excited at the treatment they received, and I think that the Red Osmia is the tamest of all the bees I have had anything to do with on my bee wall.

As to teasing the bees—or should I say making scientific experiments upon them?—the most obvious thing to do was to change the tubes so that a bee on returning home would find some other bee's tube in the place where its own ought to have been. These experiments did not seem to show much of what we should call 'intelligence' on the part of the bee. For instance, if a bee was just beginning to build the front door of a tube it had stocked and it was given an empty tube in place of its own, it usually, though not invariably, built the front door on the empty tube, although it quite obviously knew that the actual tube was not the right one. If a bee was given a tube in which a cell had been properly closed and she happened to be just going to close a cell in her own tube, she would go on adding pellets to the completed wall although she would not bring in as many pellets as she would otherwise have done. If she were given a fully stocked cell, ready to receive an egg, and she had just begun to stock a cell in the tube which was taken from her, she would continue to bring in pollen to the loaded cell although she would only bring in a few loads.

But although she shows either lack of intelligence or a very slow-working mind, she usually will complete a piece of work however long and however much trouble it may take. By removing a front door every time it was nearly finished during a period of three hours, a bee has made more than forty journeys fetching pellets to complete a job which is usually completed with about twelve pellets. After three hours I gave up and the bee finally completed the door.

Just as one can experiment with the adult bees, so one can experiment with the larvæ, and it is rather surprising to find that most of the larvæ will eat considerably more pollen paste than is provided for them. Some larvæ managed to eat twice their original supply. In the same way larvæ deprived of some of their food did not die of starvation but, most philosophically, began to spin up when they found there was nothing left to eat. The only result of these feeding experiments was, apparently, that the size of the bee was affected; that is, the larvæ from which food was taken hatched out into smaller bees and those which ate extra food tended to produce larger bees.

I experimented with different kinds of food and found that pollen paste stored by other bees was quite a suitable diet, but that various things such as cocoa, chocolate, swiss milk, bovril and virol, were of no use for nourishment, although they were readily eaten, the grubs always dying after a longer or shorter time. But I have said enough about the Red Osmia to show you how easy it is to observe its habits, and that is my object here rather than to go into details as to the habits, as will be done elsewhere.

As I have mentioned, various other bees occupied my tubes, such as the Blue Osmia (*O. ænea* (*cærulescens*)), the Leaf-cutter bee (*Megachile*), the Carder bee (*Anthidium*) and others, so that anyone who starts a bee wall is likely to have plenty of entertainment as, although one kind of bee may only be about for a few weeks, its place is taken by another, and so on, all through the season.

All the bees have the same habit of storing a mixture of pollen and honey in their cells, but the materials used in cell construction vary and the habits of the different kinds differ in various ways.

But now I want to say something about wasps which will also be found using the tubes, and we can now refer to one of the habits of wasps in which they differ very much from bees. Whereas all bees feed their offspring upon pollen paste, the wasps store up insect or spider material for their larvæ. But whereas one kind of wasp will store, let us say, caterpillars, another stores grasshoppers, another flies, and so on. Nor does the one which stores, let us say, flies store all kinds of flies, but confines itself to certain kinds, and thus we find that each kind of wasp only stores one or two kinds of material in its cells.

Whereas pollen and honey will keep good and suitable as food for the bee grubs for a very long time—I have found material two years old quite good—insect or spider material would quickly go bad if it were stored in sufficient quantity to last the wasp grub during its growing period. We therefore find that the few wasps which provide dead material for their offspring mostly bring a little at a time and return with fresh supplies at intervals during the growth of the larva, while the majority of the wasps store up living material. But if a wasp were to bring home a living spider or insect and place it in the cell, the probability is that, by the time it arrived with another one, the first would have walked out or flown away and therefore, to prevent this, the prey is paralyzed by being stung by the wasp. It can still exhibit movements, slight twitchings, bending of the body or movements of the jaws, but it cannot coordinate its movements so as to walk or fly away. I have kept alive these paralyzed spiders, caterpillars and other insects for three or four weeks, and thus the larva of the wasp feeds during its whole life upon fresh living material, each individual insect usually remaining alive until a large part of it has been eaten by the wasp grub.

Among the solitary wasps which are the commonest users of my 'wall' are the Odyneri, of which two or three different kinds are usually to be found, though each kind has its regular season (v. Plate 11). *Odynerus callosus* is a very early insect, to be seen about in March and commencing work at the bee wall in April. Just as the Osmia is dependent upon the opening of the flowers and the supply of pollen and nectar for the commencement of its cell stocking, so the Odynerus is dependent upon the appearance of caterpillars of a suitable kind and size, and the particular caterpillars which this wasp always uses are small green ones and brown ones which, if left alone, would develop into small moths about the size of a large clothes-moth and belonging to the same family. These caterpillars the wasp obtains by hunting amongst the leaves of apple and other trees and, having seized one with its jaws, the wasp bends the tip of her body beneath that of the caterpillar and stings it several times, the number of times apparently depending upon the activities of the prey. As the eggs of the moth do not hatch until the leaves of the plants upon which the caterpillars are to feed have opened out, a cold and late spring may delay the bursting of the buds and the hatching of the moth eggs and again the commencement of the work of *Odynerus callosus*, though whether the *callosus* makes up her mind as to when to commence building after going round to see how the caterpillars are getting on, I don't know!

Now this wasp differs in its cell-building from the Osmia in several ways. You will remember that the Osmia always marked out the length of the cell with one or two pellets before she commenced to stock with pollen and honey and that only after completing the store of food does she lay her egg. The Odynerus never places any marker pellets and the egg is laid either before any food has been stored or when very little has been brought in.

The *callosus* usually stores eleven or twelve caterpillars in a cell though the number actually varies from four up to thirty-two, depending partly no doubt upon the size of the caterpillars and partly upon the will of the mother, because she supplies more food in the cells in which daughters will grow up than in cells in which only sons will be produced.

I have never persuaded a *callosus* to occupy a glass tube though I can always obtain many series of cells in the elder tubes (*v.* Plate III), and by splitting these open and transferring the contents to the elder pith cells already referred to, it is easy to follow out the life history of the insect. The cell series are arranged in the tubes in the same way as those of the Red Osmia, the outer wall of one cell acting as the inner wall of the next, and I have had as many as sixteen cells in such a series in long tubes, although from two to seven is more usual.

The material used is usually of a clayey nature, a lump of marl lying near the 'bee wall' being a constant resort of Odyneri which can be seen moistening and scraping the surface and carrying away the pellets.

And this moistening of the material in order to make it into pellets is something different from what we saw in the Osmia and raises the interesting question—Where does the moisture come from? It is quite easy to see that the moisture comes from

PLATE III

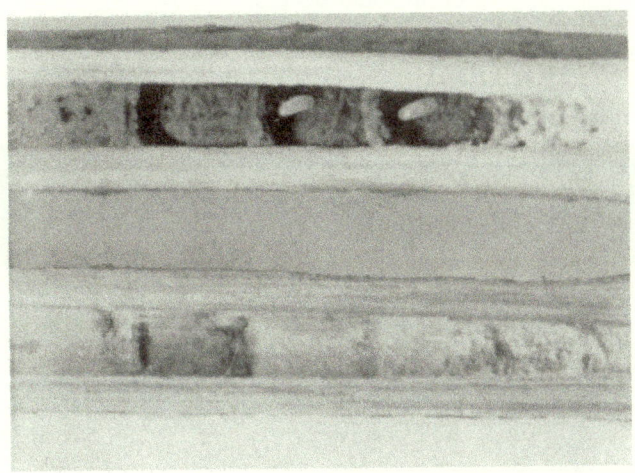

An elder tube, which has been occupied by the Red Osmia, split open and showing three cells stocked with pollen paste and two eggs in position in two lower cells. Unfortunately the egg in the third cell was destroyed in splitting the tube.

An elder tube, which has been occupied by one of the species of Odynerus (*Odynerus callosus*), split open and showing one cell. In the inner end (left) is suspended the Odynerus egg. A number of paralyzed caterpillars are seen, and just inside the cell at the outer end (right) is seen the small egg of a parasite (*Chrysis*).

the mouth of the insect and we at once imagine that it is saliva which is used. If, however, we watch the Odyneri at work we shall find them sucking up drops of dew and even the froth produced by the young cuckoo-spit insect, and if a pond is near at hand, many of the wasps will be seen along the edges drinking water. In fact, at very short intervals, these Odyneri stop their ordinary work and go away in order to renew their supply of water which is carried in the crop.

The larva, when it hatches, is a minute transparent grub of a pale yellow colour. The egg-shell splits along the side and the little grub wriggles out and falls on or amongst the paralyzed caterpillars upon one of which it at once fixes itself by driving its very fine and sharp jaws through the skin and, having thus anchored itself, it commences to suck the blood of its host.

As the larva sucks, it changes its colour according to the colour of the caterpillar upon which it is feeding. If the latter is green, the wasp grub becomes a beautiful pale transparent green but, if the caterpillar is brown, the grub becomes a transparent amber colour. If, however, a green grub transfers itself from a green to a brown caterpillar the colour of the grub first becomes muddy and, in the course of a few hours, changes to amber, the reverse being the case if an amber-coloured grub is fed upon a green caterpillar.

If we now cut open a green caterpillar we shall find that its blood is green while that of a brown caterpillar will be found to be amber, and the sacrifice of a wasp grub will show us that its colour is due to that of its blood which is visible through a transparent skin. It seems, therefore, that the blood of the prey becomes the blood of the grub, almost unchanged, or at least that the colouring material of the blood passes from the caterpillar to the grub.

The wasp larva is full grown in about eleven days, in which time it has devoured all the caterpillars provided for it and it then changes its colour and appearance, becoming pale cream coloured whatever its food has been, and spinning a white silken cocoon. In this cocoon the larva rests for a few weeks, when it changes to a pupa and soon hatches out into a wasp which again builds cells and stores them to produce a second generation of grubs. These, when full-grown, form a silken cocoon as before, but usually rest until the following spring before turning into a pupa, from which, a few weeks later, the adult insect hatches out and escapes. The Odynerus respects its sisters and brothers and will not destroy their cocoons unless they are already dead, but if it can succeed in squeezing past a living cocoon it will make its escape.

Experiment showed that the larvæ would almost always eat more than their mothers had provided for them, and a trial with various other kinds of food proved that almost any kind of insect or spider material, so long as it was soft enough, was readily eaten by them. And this raises the interesting question: If these wasp grubs will eat various other kinds of food, why do the different kinds of wasps always store up some special kind of food? The explanation is, perhaps, that the habits of the mother-wasps are to some extent due to memory of their own youth, and this suggests the possible experiment of bringing up a number of grubs of one of these wasps upon a food quite different from that supplied by their mothers and seeing whether this will influence them when they are grown up and begin to stock the cells for their own offspring. I don't suppose that the choice of food is entirely due to memory but it would be very interesting to see whether the wasp could adapt its method of attack to that necessary for dealing with a different kind of prey.

I have only outlined the life and habits of one kind of Odynerus, although several kinds are likely to be found at the bee wall. One large species, *Odynerus antilope*, is more obliging than *callosus* since it readily occupies the glass tubes, but it is a very shy insect and always ceases work and escapes as quickly as possible when any attempt is made to see it in the tube by withdrawing the latter from its paper sheath. It only has one generation in the year and thus differs greatly from the *callosus* which, on rare occasions, may produce three and always produces two.

There is a third species which, although not a frequenter of the bee wall, is not rare at the clay box already mentioned. This is a somewhat smaller insect known as *Odynerus spinipes* and its interest lies in the fact that, while the other two species mentioned store up caterpillars, *spinipes* stores up small beetle larvæ which it finds upon vetches. Secondly, it constructs at the entrance to the burrow an elaborate structure which projects from the face of the bank or clay wall in which the burrow is made, and thirdly, this insect very commonly makes new burrows for itself instead of using old ones, the other two species mentioned much preferring to occupy ready-made burrows.

These elaborate gargoyle-like 'porches' are frequently washed away by rain but the insects renew them again so long as they are still working the burrow, because the porches are not constructed of pellets brought from outside but always, so far as I know, of pellets brought from inside. Thus, if a porch is removed while the owner of the burrow is collecting beetle larvæ to store a cell, the renewal will not take place until another cell is commenced, and here again we get another difference in habit from the other two species. *Spinipes* does not usually construct its cells in series but separate cells are cut out from the sides of the tunnel and these vary in direction according to the conditions of working. Thus, although horizontal cells are usual, slanting or even perpendicular cells are made if, for instance, a stone or an extra hard spot prevents the normal direction being taken.

I have, in this lecture, outlined the life histories

of two main types, one a bee and the other a wasp, and have told you the methods I employed to discover their life histories. Further, I have mentioned a few other types in order to show you that each kind has its own peculiarities. It is these peculiarities which enable different kinds to live together in the same locality and to be members of a community, not all competing for the same food and the same spot, but fitting in with one another in a most complex manner. We see, therefore, that life history work may help to explain the relationships of species in a community.

Lecture III

THE HABITS OF CATERPILLARS

WHETHER or not we are interested in insects there are few people who do not know that a caterpillar ultimately becomes a butterfly or moth. It is true that some people believe that it is a kind of worm and it has been more elaborately described as an 'upholstered worm,' but most people know, in outline, the life history of the butterfly with its four stages—egg, caterpillar or larva, chrysalis or pupa and perfect insect.

Now although most people can recognise a caterpillar, they are very apt to recognise as caterpillars larvæ which are not the young of butterfly or moth, since the larvæ of the sawflies—relations of the bees and wasps—which have the same habits as caterpillars also have a general resemblance to them.

One of the points of interest about a caterpillar is its number of legs, and some of you may remember the difficulties into which a certain bishop got upon the question 'How many legs has a caterpillar got?' as recounted in a poem.

As a matter of fact a caterpillar has only six true legs—and no insect has more—but it also has a certain number of 'prolegs,' never more than five pairs and sometimes not more than two pairs, but this fact is of importance in distinguishing it from the caterpillar-like larva of the sawfly. The sawfly larva has more than five pairs of prolegs and these begin on the segment next behind the ones bearing the true legs, a segment upon which the caterpillar never has any prolegs.

The caterpillar, as everyone knows, hatches from an egg and it is the growing stage in the life history of the butterfly or moth. It is a voracious feeder and, according to its kind, it feeds upon leaves, buds, fruits, roots or even on wood, and there are even caterpillars which feed upon fur, feathers and horn.

Most caterpillars produce silk and it is really with regard to the uses that they make of this silk that I propose to devote this lecture. Silk, as produced by the caterpillar, is the product of the salivary glands which produce a liquid which possesses the peculiar property of becoming solid in the air. Projecting from the lower lip of the caterpillar is a short tube, the 'spinneret,' which is really the tongue and the silk flows from a minute hole in the end of the spinneret.

The sawfly larvæ already referred to, caddis larvæ—the so-called caddis worms—ant, bee and wasp grubs and various other insects produce silk in the same way, but certain other insects produce silk from other glands and, as you probably know, spiders have their spinnerets near to the posterior end of the body. Thus silk, although always produced in the body as a fluid which solidifies on contact with air, is not always produced by the same glands.

Now the caterpillar uses its silk in various ways of which the best known is the formation of the cocoon in which the full-grown caterpillar changes to a pupa. Not all caterpillars form a cocoon and the amount and quality of the silk produced differs profoundly in different kinds. In some cases only one or two threads are formed, which bind the caterpillar to its support before it becomes a pupa and later serve to hold the pupa in position until the butterfly emerges. Very few kinds of caterpillars, during the spinning of the cocoon, produce a continuous thread of sufficient strength to be reeled off and used by man, and of these the true silkworm has been cultivated for so long that it is now really a domestic animal, and the moth seldom, if ever, uses its wings. Other kinds of silkworms form cocoons of a quality which, although not suitable for reeling, can be spun in the same way as cotton fibres are spun, but very few caterpillars produce silk which is of any use to man, although it is always of considerable use to the caterpillar and it is of the uses of silk to the caterpillar that I wish to speak now.

All silk-producing caterpillars are perpetually dribbling so long as they are not resting—to use an expression more familiar in connection with a teething baby—and as the caterpillar moves forward, the saliva, flowing from the spinneret, is laid down as a thin thread, the legs and prolegs of the moving larva passing on either side of it. While the caterpillar feeds, the oozing saliva is swallowed with the food and assists in its digestion.

Now the leaf-feeding caterpillar is always in danger of falling from the leaf or branch where it is feeding. A gust of wind may blow it off or a bird, suddenly alighting in the neighbourhood, may shake it off, and so on. This may entail a long and weary walk back to the feeding place, but the perpetually oozing saliva now comes into play. As the caterpillar falls, the saliva runs out from the spinneret as a thread which forms a life-line and up which the caterpillar can climb back to the place from which it fell. A wonderful brake-apparatus exists just inside the spinneret and, by exerting certain muscles, the caterpillar can press down a horny rod upon the glutinous thread and thus control the outpouring of the silk. To climb up, the insect uses its jaws for holding on to the thread which it slowly twists round its legs and, having recovered its position, it throws away this tangled mass of silk.

When the caterpillar rests it stops the flow of saliva but, previous to resting, it takes certain precautions. It spins a number of threads on the support where it rests and holds on to this 'mat' by means of the numerous fine hooks at the ends of the 'prolegs.'

Caterpillars, while growing, cast their skins a number of times and each time, before so doing, the silk-spinning caterpillar lays down a silken mat on the leaf or branch, to give itself a firm hold.

We have now seen three uses to which the caterpillar puts its silk. It uses it as a life-line, as a resting- or moulting-mat and as a cocoon or binder at the time when the pupal change is about to take place. But there is another use for the silk and that is in forming or helping to form a shelter in which the caterpillar can rest or inside or under which it can feed. Such a shelter varies from a crack under the bark lined with silk to a curled leaf tied up with silk or a 'case' composed of various materials tied together with silk, capable of being transported about, the caterpillar retaining it round its body in the form of a tube from which it projects its head and leg-bearing segments when it wishes to move but into which it can withdraw its whole body. The structure and form of these movable cases varies greatly and a few examples will be useful.

A familiar one, perhaps, is that of one of the clothes-moths whose caterpillars feed upon various clothing and other household materials, such as carpets and curtains and even leather goods. In this example, the small caterpillar, as soon as it hatches, bites off small fragments of the food material amongst which the eggs were laid by the foreseeing mother and, fastening them together with silk, forms a minute tube into which it creeps and which it perpetually carries about with it, adding to the front end as the necessity for greater length and diameter arises and biting off the smaller end when it becomes unnecessary.

When the caterpillar wishes to rest or to change its skin, it withdraws into its case, having first anchored it to the support by means of one or two threads and, on resuming active life, it bites through these threads and carries its case away.

A related moth, known as the Larch case-bearer, forms its case out of the apex of a larch needle, lining it with silk. When this little caterpillar feeds, it bores a hole with its jaws in a fresh larch needle and ties the case with silk so that the opening of the case fits over the hole in the needle and it then feeds upon the tissue within the needle, gradually stretching farther and farther out of its case as the food supply becomes exhausted until only its tail end remains within the case. It then draws back and begins to attack the pine needle in the opposite direction and thus it hollows out an area of the leaf on either side of the entrance hole and the leaf turns white and withers. When the caterpillar can get no more food out of the boring it breaks the connection between its case and the leaf and moves elsewhere. During one season's feeding this little caterpillar only grows sufficiently to fill the little case it had made for itself out of the end of a larch needle but, after sleeping out the winter inside the case, firmly fastened by silk in a crack in the bark of the tree, it finds the necessity for enlarging the case, which it does by splitting it up the side and adding a 'gusset' in the gap.

A more elaborate and, in many cases, a much larger 'house' is that made by the so-called 'Bagworms,' caterpillars of the Psychid moths. Our British kinds are small and inconspicuous but in many parts of the world the Psychid cases run up to 3 inches in length. These cases are composed of various materials according to the habits of the larva: bits of stick, leaf stalks, lichens or grass, these being fastened together with silk and the whole structure being lined with the same material. In some kinds the shape of the case is very peculiar and there are some in the British Museum which were sent there in the belief that they were the shells of snails.

These Psychid moths might be described as a degenerate race and their degeneration is in some way associated with the case-making habit of their caterpillars. The female is no longer what we should recognise as a moth but is grub-like in form, possessing no wings and being so helpless that she never leaves the 'house' which she had constructed in her caterpillar days. She is so degenerate, in fact, that in most cases she cannot even lay her eggs and she dies with the eggs still in her body, so that the young caterpillars, when they hatch out, have to bite their way out of their dead mother before they are free.

In Brazil the caterpillars of the Hammock moth make a strange shelter in which they live and which they carry about with them, enlarging it as they grow and ultimately using it as a protective cocoon in which to change into the pupa stage. This structure is composed of the excrement pellets of the caterpillar bound together with silk and these pellets, as if specially shaped for the purpose, are flattened and thus different from those of other caterpillars. The case has both ends alike and looks something like a gondola; it is open at both ends and the caterpillar rests either way inside it, its head only being visible projecting from the opening.

As the caterpillar grows it enlarges its case by adding material at either end. The case is always attached by three or four strong silken cables to the leaves and twigs amongst which the caterpillar is feeding, each of these cables being composed of many threads spun backwards and forwards twenty or thirty times. In changing its position an old cable is never cut until a new one has been made. In feeding, the caterpillar extends out from the case and looks like a thin ribbon and, having seized a leaf, it cuts it from the plant and then partially withdraws into the case. There it feeds, holding the leaf with its feet so as to steady it.

When it is ready to pupate, the caterpillar, which so far has always kept its cradle more or less horizontal, cuts the cables at one end so that the structure hangs perpendicularly and, having added more silken threads to secure the case in this position, it retires within, closes the entrances with silk and changes to a pupa.

This anchoring of the case by the Hammock

caterpillar has of course made things a little more difficult for it and has involved more work—stretching to its full length from the case to reach more distant food, the building of the strong cables and their frequent renewal when the necessity for moving on arises—and we are thus being introduced to a new aspect of this house-building habit of the caterpillar, when conditions are such that the house cannot be moved and journeys have to be made from it to obtain food, and back to it for shelter. How this habit has arisen it is difficult to say, but it is largely associated with another habit, the living together of members of a family. One might perhaps suggest that when several members of a family made themselves shelters intending to carry them about and, being close together and liking one another's company, happened to fasten them together, the necessity for adopting the habit of leaving the home and going out for meals arose from the fact that by all being pulled in different directions, the homes refused to go with their owners.

However, before talking about these fixed homes it will be interesting to notice that, in other Orders of insects, this case- or shelter-making habit has arisen quite independently. For instance, the so-called caddis 'flies' have larvæ showing a slight resemblance to caterpillars and many of these make cases which they carry about with them. These larvæ, however, live in the water and make their cases of materials similar to those used by the Psychid caterpillars, except that they obtain everything in the water. The case is lined with silk and fastened together with silk, and the larva has special means of holding on inside while its head and leg-bearing segments are protruded from the front end.

In the Order to which ants, bees and wasps belong some of the sawflies are case-makers, but they differ somewhat in their habits from the caterpillars and caddis larvæ since, while these two types finally use their cases as protections for the pupa, the sawfly larva leaves its case when full-grown and descends into the ground where it spins a silken cocoon.

We see, therefore, that the case-making habit is not a peculiarity of the caterpillar but has appeared in at least three different Orders of insects. But even amongst butterflies and moths it has arisen a number of times in quite independent families as the examples previously given will serve to show.

Now caterpillars are not what one would usually call 'active' creatures and although many kinds exist which are to be found living independent lives, such that we may have a difficulty in finding more than one or two on a plant, there are other kinds which live in parties and the life histories and habits of these are interesting.

A great many butterflies and moths lay their eggs in masses and as, in most cases, the eggs are laid upon the plant which will afford food to

the caterpillar, the latter finds its food available as
soon as it hatches from the egg. But a crowd of
caterpillars will soon exhaust the leafage in a particular spot and this will cause the individuals to
disperse in search of other feeding grounds so that,
in this way, a family very quickly becomes broken
up. In many cases, however, although the family
may not remain together as a whole, the individuals,
for a shorter or longer time, show a taste for one
another's company and can be found in larger or
smaller batches upon the food plant.

If we study the habits of such a moth as the
Emperor (*Saturnia pavonia-minor*), a species not
uncommon in the British Islands, we find that
the female lays her eggs in one or more batches—
there may be two or three hundred in such a batch
—upon various food plants such as meadow-sweet,
heath, willow, blackthorn, apple and various other
species. The larvæ, when they hatch, move about
over the leaves, always with the spinneret oozing
out saliva so that silken threads are laid wherever
they move. The silken mat thus laid down by all
the caterpillars gives them a foothold on the foliage
and they can be seen in their black coats—for it
is not until later that they acquire their beautiful
green colouring—feeding in little parties of three
or four or perhaps twenty. But the parties break
up after a short time and each caterpillar is there-
after a solitary insect leading its own life without
any regard for its brothers or sisters. Such an
insect however shows us a very elementary type

of social life which endures for a short time, social life amongst insects being definable as life in which the individuals contribute something for the common good of the family. In this case all that is contributed is a silken carpet, the joint work of many caterpillars.

The Buff-tip moth (*Pygæra* (*Phalera*) *bucephala*) lays her eggs in batches of from thirty to sixty on the upper side of the leaf of elm, lime, hazel or other tree and the young caterpillars, on emerging, feed in company, at first only skeletonizing the leaves but later eating everything except the main veins. The parties gradually move to the tops of the trees where their presence can be detected by the damage they do, long bare twigs, perhaps several together, standing out amongst the foliage.

Examination of the bared twigs shows that they are covered with fine silken threads which, spun by all the members of the party, have given each of them a foothold. The parties move about rather freely, changing their feeding grounds and descending to a more sheltered spot on a windy day so that they gradually become broken up, although individuals still enjoy one another's company and, even when full grown, two happening to meet will tend to stay close together. One interesting point to notice is that while the parties are still in existence the caterpillars all moult together, so that numbers of shrivelled skins may be seen attached to the thin mat of silk at some particular spot amongst the damaged leaves.

The full-grown caterpillars lose interest in one another and become very restless, descending from the food plants and wandering over the ground where they ultimately change into pupæ in a tuft of grass or under fallen leaves, without making any cocoon.

The caterpillars of the Large White or Cabbage butterfly (*Pieris brassicæ*) of our gardens are another example of the same kind. As before, the eggs are laid in batches of a few up to twenty or more, on the under side of the leaf of the food plant which, in this case, is a low growing plant such as cabbage or cauliflower, or Tropæolum (the so-called Nasturtium). The young caterpillars keep together, spinning a silken mat which is often very conspicuous—depending of course upon the number of individuals contributing. Even though they may disperse while feeding they usually collect again in parties to rest and, not infrequently, caterpillars of different sizes become mixed in these parties, showing their desire for company. As in the case of the previously mentioned species, the full-grown caterpillars lose this herding instinct and disperse to pupate, the pupæ being familiar objects upon walls, under the eaves of sheds and in various other places. A near relation of this butterfly, the Small White (*Pieris rapæ*), lays its eggs singly in the same situations as does the Large White, but its caterpillars show none of the gregarious instincts of the other (*v*. Plate IV, fig. I).

In the Small Tortoiseshell butterfly (*Vanessa*

PLATE IV

The Carpet Web. Part of a family of caterpillars of the Large White butterfly (*Pieris brassicæ*) resting upon the common web spun by them all. With them is a single caterpillar of the Small White butterfly (*Pieris rapæ*) whose territory they have invaded.

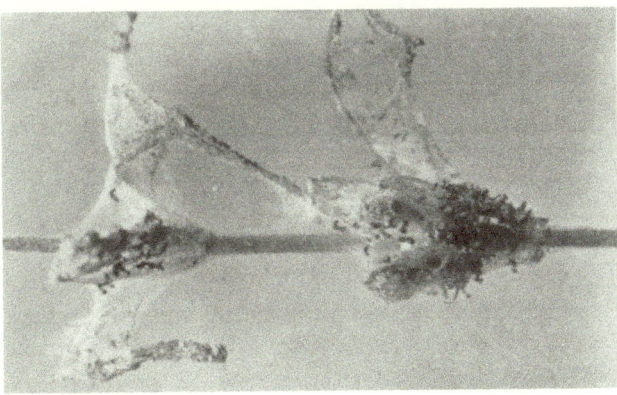

The Feeding Web. Part of the web of the caterpillars of the Peacock butterfly (*Vanessa io*) on the nettle, showing the moulted skins on the outside. After moulting on the web, the caterpillars migrate elsewhere.

urticæ) and the Peacock (*Vanessa io*) we see some interesting habits. The nettle is the only food plant and the eggs are laid in masses, usually near the top of the stalk. The young caterpillars herd together, the whole family feeding together, but here we find something different from what we have seen in the other kinds mentioned. The silken threads spun by these little caterpillars, instead of covering the surface of the plant as a carpet to provide foothold, are so spun that they form a shelter within which the caterpillars feed. During the first few days their united appetites are not such as to cause them to move far from the spot at which they have hatched, but as their hunger increases they move as an army, spreading their silken tents wherever they go and quickly skeleton-izing the leaves and webbing the flower stalks so that the nettle-head soon assumes an appear-ance as if it had been tied up in spider-webs. The family changes its skins in parties and, after the moult, the party will perhaps move across to another stalk and climb to or near the top before the individuals again commence to feed (*v*. Plate IV, fig. 2).

After the second or third moult, the web-spinning activity of these caterpillars is less notice-able, the family becomes more and more broken up into small parties, and the parties of two to perhaps five—or even odd individuals—now tend to make homes for themselves by closing up a nettle leaf, tying the edges together with silk.

Having succeeded in this, the leaf is eaten away and the caterpillars then proceed to do the same thing again. The silk-production of the larva gradually diminishes and, as it becomes full-grown, it loses its interest in its relations and ultimately, without any regard to the proximity of others, suspends itself head downwards from a nettle leaf and there changes into a pupa.

One interesting point in which these caterpillars differ from those previously mentioned is in their habit of wriggling and dropping to the ground if touched or often even if a shadow passes across them. This we can explain by the fact that whereas tree-feeding caterpillars and those which feed in areas where many kinds of plants grow mixed together may have considerable difficulty in getting back to their food plant if they fall from it, no such difficulty is likely to occur where the food plant occurs in a mass, as the nettle does.

To pass on to another example quite unrelated to the butterflies last mentioned we find that the caterpillars of a small moth—the Little Ermine (*Hyponomeuta* or *Yponomeuta*)—show the same habit of spreading a silken web all over the food plant, which is usually the common spindle-tree (Euonymus), but in this case a number of families are frequently concerned in the same web, apparently owing to the sluggishness of the parents who have not troubled to leave the spot where they themselves grew up. Occasionally long stretches of hedgerow are hung with the fine webs of this caterpillar so that the place looks as if fine muslin had been hung out to dry.

In all the examples so far mentioned the silk has covered the food plant in the area where the caterpillars feed, but we can regard those kinds in which the silk forms shelter webs as having attained a higher stage in social life than those by which only matting is laid down. There are, however, a number of kinds of moth and butterfly in which the web, instead of being spread as a shelter over the whole feeding area, is limited in size and forms a more or less permanent home from which the caterpillars go out to feed and to which they return to rest. As will be seen later on, this is only an advance on the last-mentioned type, as more or less intermediate types exist between the two.

As a first example of this limited web I will take the Lackey moth (*Clisiocampa neustria*), so called because of the colouring of the caterpillar which reminds one of the livery of the flunkey of the Victorian age and earlier. The moth lays her eggs in a spiral, round a twig of apple, plum or other food plant, covering them with a varnish which protects them from the rain. The egg-mass remains on the twig throughout the winter and in the spring the young larvæ hatch, when they generally move out to the end of the twig with its now opening buds. Here they construct a small web by binding together the young leaves and in this web they feed. In about a week they assemble together on the outside of this little web and cast their skins, after which they move off elsewhere and commence the construction of another web. The journey after the first moult may be a long one. If the eggs have been laid upon one of the lower branches of the tree, as is usual, the second web may be constructed near the top of the tree and perhaps on the other side, 15 feet or more from the birth-place of the caterpillars.

In one case a first web was built upon the shady side of a hawthorn hedge while the second was more than 12 yards distant on the opposite side of the hedge which was very dense and about 5 feet in width. This second web, built at first round leaves which serve as food, is the residence of the family for the next two or three weeks and, although the food supply within it is quickly exhausted, it serves as a meeting-place and a resting-place to which the individuals return again and again. As each caterpillar spins a silken thread as it walks, the various roads to the feeding grounds around the nest soon become white with the numerous threads laid down upon them and, as is obvious, the caterpillars can always find their way home by following these silken highways, along which they are usually to be seen, hurrying to or from the feeding grounds. But although there are times when the majority of the family seem to be either in or out, there appear to be no regular hours nor any particular weather-conditions which govern the movements—except perhaps that in hot sunshine they may rest on the shady side of the nest instead of on the top of it.

As the Lackey caterpillars grow they seem to get very careless in their movements and it is not at all an uncommon thing to see one moving rapidly along a branch and then falling off the end of it. The silk which they produce is not of sufficient strength to bear their weight so that the caterpillar, once off familiar ground, has not much chance of returning to it and is very unlikely, when it falls to the ground, to find its own tree again. Therefore, by the time that the third web is made, the family is usually greatly reduced and the remnant, now well grown, moves down the tree to one of the larger forks where a fairly dense mat is laid down in a very exposed position, upon which the individuals rest on their return from feeding, if rest it can be called, since most of the time they spend in rapidly wagging the front half of the body from side to side at short intervals, only the hind part of the body being motionless—a movement made, perhaps, with the object of driving off parasitic insects. As a fact, this final mat-web is frequently not made because the family has become completely dispersed before the time comes for its construction. The full-grown caterpillars disperse and each spins for itself a white silken cocoon amongst the leaves of the tree, in which the change to a pupa takes place and from which, a month later, the moth emerges.

Another interesting caterpillar, not unlike the Lackey in its habits, is the Processionary caterpillar (*Cnethocampa processionea*), whose habits have been described by Fabre in his *Souvenirs Entomo-*

logiques. It is a common insect in the pine-woods on the continent, and I first made its acquaintance at Vigo, in Spain, as it does not occur in the British Islands.

The female is a very heavy moth and unwilling to fly far, so that she usually lays her eggs upon the pine needles on the lowermost branches of the trees. The eggs are laid in masses and covered with flat scale-like hairs from the apex of the body of the mother so that they are disguised, the scales serving to conceal them. The larvæ, on hatching, spin a small web amongst the needles in the neighbourhood of the egg-mass and they feed within this and thus gradually destroy its supports. When the nest collapses the larvæ move out in procession, one leading, the next one with its head just touching the tail of the first, and so on, and they construct a new shelter. During the autumn several of these webs are constructed, each one being usually higher up the tree than the last so that, on the approach of winter, the caterpillars have usually arrived somewhere near the top, whether it is ten, thirty or more feet in height.

As the cold weather approaches they change their feeding habits and cease from destroying the rafters of the nest. They spend most of the day, if the weather is fine, resting upon the upper surface of the web and expanding it, the whole family of three hundred or more sallying forth about dusk and breaking up into small parties and feeding all round about, but returning to the web before morning. By early December the nest may be 6 inches or more across and this structure forms the home of the family all through the winter and until the following spring when the caterpillars disperse to bury themselves in the soil in order to pupate. All through the winter and early spring, on suitable days, these caterpillars make journeys, sometimes to the ground and many yards from the home tree, formed up in procession and led by an individual who apparently has no better claim to leadership than the fact that he is in front. Any individual may become leader, as can easily be shown by breaking up a procession.

Fabre recounts how he induced such a procession to climb a large palm vase nearly a yard and a half in circumference at the top and, having got the line to walk round this circle, he broke off the procession in such a way that the last caterpillar to reach the top fitted into a complete circle formed by the others, the leader touching the tail of this last of the column. By this means Fabre, unbeknown to the caterpillars, had done away with the leadership and the procession marched round and round the vase for several days and in the end apparently only escaped by chance from the enchanted circle. Each caterpillar in the procession lays down its silken thread so that the column produces a well-marked highway along which it is easy to retrace its steps on the homeward journey.

The next example is a British one—the caterpillar of the Small Eggar moth (*Eriogaster lane-*

stris), an insect which, although occurring throughout Britain and in Ireland, is fairly common round
about Cambridge. In this case the female lays her
eggs in a spiral round a twig of hawthorn or blackthorn and, as she does so, she sheds the long furry
hairs which form a tuft at the posterior end of her
body so that the egg-mass is smothered in dark
brown fur (*v.* Pl. v, fig. 1). The eggs, like those of
the Lackey, remain throughout the winter and in the
spring, soon after the buds have opened, the caterpillars bite their way out and at once begin to form
a web round the egg-mass, eating up any leafage
which becomes included in the nest. And this insect differs from those previously mentioned in that
only one nest is constructed and it is perpetually
added to as the caterpillars grow, so that it becomes
very large and, being white and usually upon the
face of a hedge, it is very conspicuous. Such a
nest may contain three or four hundred caterpillars and its visibility is increased by the bareness
of the hedge round about where the leaves have
been destroyed by the voracious family and by the
branches carrying the white silken roadways (*v.*
Pl. v, fig. 2 and Pl. vi, figs. 1 and 2).

If such a nest be examined certain interesting
points will be noticed. First, there are usually two
or three holes in the outer layer in the upper region
and these are the normal passage-ways in and out,
though a caterpillar inside and wishing to get out
will not infrequently bite its way through where
it happens to be, thus creating another opening
which may or may not be allowed to remain.

PLATE V

The Small Eggar moth (*Eriogaster lanestris*) laying her eggs and covering them with brown velvety fluff from the apex of her body.

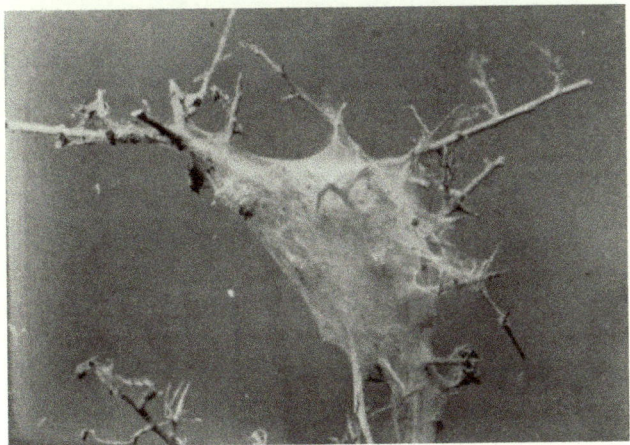

A silken nest of the Small Eggar caterpillars (*Eriogaster lanestris*).

PLATE VI

The Home Web. A small nest of the Small Eggar caterpillars (*Eriogaster lanestris*) with the caterpillars resting upon it.

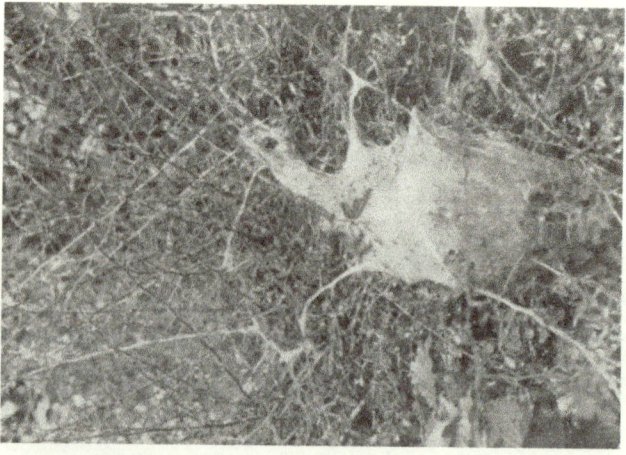

The Home Web. A large nest of Eggar caterpillars showing the silken highways running out in all directions to the feeding grounds. Notice that all the leaves of the hawthorn in the neighbourhood of the nest have been eaten.

Examination of the nest through one of the openings will reveal the fact that close beneath the outer silken layer is another which may or may not have an opening visible from this view-point, but if so, another layer will be seen inside the second. If, however, the nest be examined by removal of layers at one side, it will be found that the additions are seldom if ever complete envelopes but merely additional chambers, added chiefly to the sides so that the majority of the partitions in the nest run more or less perpendicularly.

Now this apparently trivial detail has an important bearing upon the life of the family because, owing to this, the dry excretory pellets dropped by the caterpillars inside the nest tend to fall to the bottom where they accumulate in a part of the web which first sags with their weight and often ultimately gives way so that, towards the end of the nesting period, a large proportion of the frass has fallen out of the nest.

The caterpillars, hatching in April or May, are full-grown by the end of June, shortly before which time they tend to wander farther and farther from home so that the numbers gradually diminish. I have found these caterpillars travelling at full speed on the ground more than 40 yards away from the nest, having passed through a narrow border and across a tennis lawn. They ultimately build a blunt-ended oval cocoon of a consistency not unlike egg-shell and of a colour which, Prof. Poulton has shown, varies with that of the background upon which the caterpillar spins it. The pupal stage lasts in some cases only until the following September or October and in others continues for two or three years, no explanation of this long delay in the emergence of the moth having so far been suggested[1].

In all these cases, with the not infrequent exception of the Little Ermine moth (*Yponomeuta*), the caterpillars disperse when they are full-grown, but there are many kinds of moths and some of butterflies in which the family remains to pupate in the common web. The South African Silk moth (*Anaphe*) is one such example, the web being very thick and in the end becoming filled up with the cocoons of the pupating caterpillars. The caterpillars of a Mexican butterfly (*Eucheira*), related to our own Cabbage White, make a web which is almost like parchment and, when full-grown, they suspend themselves head downwards on the inside of this web and, without making any cocoon, change into pupæ.

It will perhaps have been noticed that in this lecture I have used a different method of dealing with my subject from that which I used in the previous one. There I laid stress upon the methods I adopted in order to observe the habits of a few bees and wasps. Here I have spoken about habits without any reference to the methods by which I

---

[1] Although I find, from my notes, that I have had moths emerging in the autumn, a batch of cocoons spun in 1924 has been hatching out since February (1925), no moths having appeared before February 15.

observed them. And the reason is that different insects require different methods of observation and the methods used for watching bees and wasps are quite useless for watching caterpillars. It is true that one can provide suitable food plants to receive the eggs just as one can provide bricks and tubes for the bees and wasps, but very little success rewards this method and none at all so far as any of the insects mentioned in this lecture are concerned.

Of course, it is possible to rear caterpillars in muslin cages and even in cardboard boxes with muslin tops, but the conditions are so artificial that one could not rely upon the observed habits being natural ones, and as a fact, Emperor moth caterpillars reared in cages behave very differently from those reared under natural conditions. My observations upon caterpillars have therefore been largely made in orchards and hedgerows, but where possible I have brought home the young larvæ and placed them upon food plants growing in my garden. Fortunately, I have a hawthorn hedge along one side which provides food for a number of kinds and, much to the annoyance of my gardener, I carefully preserve some patches of nettles for the Tortoiseshells and Peacocks and some wild mullein for the Mullein caterpillars, where they can live under natural conditions without being enclosed in any way. As I have said these 'desirable residences' do not attract the moths and butterflies, with the exception of the Mullein, but the Mullein moth caterpillars do not concern us here. The probable reason is that the hawthorn hedge and the nettles, although receiving a large amount of sunshine, are somewhat shut in by other plants and are therefore possibly out of the range of flight of the insects I wish to study, but when these insects, in the youngest caterpillar state or even as eggs, are introduced by me, they thrive and behave exactly as do the others left in their natal surroundings.

Although it is not usually possible to make experiments upon insects upon other people's land, I have been able to play a few tricks upon those in my garden and one point I was anxious to investigate was connected with the nest-building capabilities of the Eggar caterpillars. As I have said, they commence their nest around the egg-mass, and this structure, frequently added to as necessity demands, serves them as a home until they are full-grown. Now in many cases an insect can do something once in its lifetime but only once, and I was interested to know whether the Eggar caterpillars could re-build a nest if the structure was destroyed. I therefore brought home a batch of about one hundred caterpillars about a fortnight old and having also brought home a nest, I stripped off all the webbing from the branch, leaving only the egg-mass intact, and this branch I then tied to a branch upon my hedge and placed upon it all the caterpillars. These soon divided themselves into two parties, one of which quickly reconstructed the nest over the old centre while the other, selecting a place upon the living hedge, built a nest without having any egg-mass as a centre.

It is of course possible that the necessity for moving to a new spot may occur from time to time under natural conditions. For instance, a nest might be destroyed by hail or it might be eaten out by a hungry cow and the escaping members of the family, still retaining the herd instinct, might crowd together at some new centre and commence a new home. It is therefore unlikely that my experiment was a brand-new experience for the species, but even if it were, the habit of making a series of new nests, one after the other, occurs amongst relatives, such as the Lackey caterpillar, and is almost certainly a more primitive habit, one which belonged to the ancestors of the Small Eggar caterpillars.

By experimenting with well grown caterpillars, I found that they would reconstruct some kind of a nest amongst the ruins of the old web if the latter were destroyed so that no shelter remained. The repairs, in such a case, were by no means artistic, but they served the purpose for which they were undertaken.

I tried the experiment of slitting open a nest by cutting a short gash in the side, with the object of seeing whether proper repairs would be carried out, but here I was disappointed. No attention was paid to the gap and it was only in the course of general work that a small part of it was repaired, the wall later on becoming enclosed by a new one.

I gather from the results of these experiments that, somehow or other, these caterpillars have acquired the capability of dealing with emergencies which may occur under natural conditions, but that they show nothing which could be described as intelligence in the ordinary sense of the term.

I commenced this lecture by referring to the uses made of silk by the caterpillar, and in discussing the production of shelters made by caterpillars which herd together, I have endeavoured to trace, by means of a few examples out of very many available, the development of the structure, first as a mere carpet, then as a general web spread indefinitely over the food material, and then as a definite fixed home from which the inhabitants move forth to feed and to which they return to rest.

In conclusion, I want to point out to you that this story has been constructed upon great numbers of observations of caterpillars which were made, not for the purpose of arranging them into a story, but because the habits of different caterpillars were interesting to observe. We see, therefore, that just as mere collecting led us to speculate upon problems of geographical distribution, mere observation of habits may lead us to speculate as to the evolution of those habits, a result which will amply justify our curiosity.

Lecture IV

THE HABITS OF THE DRAGONFLY

THE dragonfly is a familiar object to anyone who lives near the marshes or the river or who has any interest in pond life. During the summer months, the large kinds may be seen hawking to and fro across the water or sometimes far away from it in some quiet country lane, suddenly pouncing upon some hapless insect and devouring it there and then, or settling upon some gate-post or other support to feed at leisure.

The smaller kinds, some blue, some red, flit about the edges of the pond or stream and seldom move far from the water, as they have not the powers of flight possessed by their larger and stronger relations.

One would perhaps expect from this feebleness of flight of the thin-bodied kinds of dragonflies that they would be much less widely distributed about the world than the thick-bodied kinds and yet, as a matter of fact, they do not seem to have suffered in this respect and are found upon isolated islands just as are the stronger relations—in fact, powers of flight seem to have had but little to do with the range of distribution of dragonflies, and this is true also of other insects. This observation has been made before and has even been immortalized in poetry:

Dragonflies and other flies are strong upon their wings
And can get almost anywhere, but there are other things
Which haven't any wings at all and yet enjoy the fun
Of getting anywhere they like, the little flea is one.
And other little creatures that hop and crawl around
Can cross the mighty oceans to wherever they are found,
But how they get from here to there, wherever they may be,
Is a scientific problem that is much too hard for me.

There is an idea amongst country people in all parts of the world that dragonflies are capable of stinging and, in English-speaking countries, they are frequently known as 'horse-stingers.' In Australia, Dr Tillyard collected evidence 'on the solemn word of various settlers in Queensland' that not only horses and cows had been killed by them but even human beings had been attacked[1]. As a matter of fact, they are quite harmless insects which, however, will defend themselves if attacked by means of the jaws, and their bite can be felt, and they also tend to bend the apex of the long body round as if they were going to sting. As, however, they have no such structure, no harm can result from this movement, unless, perhaps, the points of the piercing ovipositor of some kinds may prick the skin.

We are in the habit of speaking of dragonflies as water-insects, although the stage of which we

[1] Tillyard, R. J. *The Biology of Dragonflies*, p. 335. Camb. Univ. Press, 1917.

have been speaking occupies the air, and it will be
well therefore to define the term 'water-insect,'
which is a name given to any insect which spends
some part of its life in the water. There is no
insect which spends the whole of its life in the
water and, whatever may have been the remote
ancestry of insects, all present-day water-insects
are the descendants of land forms which have
become adapted to a life in the water and the
adaptation has been less complete in some and
more complete in others, but never wholly so. The
dragonfly is better adapted to an aquatic life than
most other insects, perhaps partly because it is a
very ancient form of insect and has therefore had
a longer time than many others to become suited
to the conditions, and this antiquity of the dragonfly is proved by the fact that fossil remains of these
insects are found in some of the oldest fossil-
bearing rocks.

Yet in spite of its long ancestry and its adaptation to life in the water, it shows that in a number
of ways it is more simple in its structure than
many other insects and even its life history has
never attained the complexity of that of other
forms, such as butterflies and moths, bees and
wasps which, speaking geologically, are much
younger than the dragonflies.

The largest dragonflies of to-day are less than
6 inches across the wings and yet fossil records
tell us that at one time the ancestors of modern
forms measured 2 feet across the wings so that,

during the ages, dragonflies have become smaller. But this is not really a peculiarity of dragonflies. We know, from fossil evidence, that horses were once much smaller than they are to-day, one of the ancestral forms having been only about 18 inches in height. We know also that elephants are smaller than they were in the past, the mammoth, for instance, having been considerably larger than our Indian or African species of to-day.

I have already spoken of large and small dragonflies, and there are in reality two groups, easily recognisable from one another upon certain characters. The large forms have been called the 'thick-bodied' as distinct from the smaller 'thin-bodied' dragonflies, but the real distinction is easily seen in the wings. The 'thick-bodied' forms have the hind wing rather different in shape from the front wing whereas the 'thin-bodied' ones have the fore and hind wings almost alike in shape. There is also a marked difference between the young or 'nymphs' of the two groups, but I shall refer to that later on.

Dragonflies differ as to their egg-laying habits. Some of the thick-bodied forms fly over the surface of the water, touching it at short intervals with the tip of the abdomen and each time this occurs a few eggs are released, which slowly sink to the mud at the bottom. These eggs are covered with a secretion which swells and becomes sticky upon contact with the water so that the eggs may adhere to water-plants and other things as they sink and, in any case, they quickly become covered with small particles of substances in the water, so that they are protected from the sight of hungry enemies. Some others of the thick-bodied forms rest upon some stick or plant projecting from the water and, submerging their long bodies, they lay upon the water-plants a mass of eggs all glued together either in a globular or worm-like form. The water acts upon the 'glue' which swells up and becomes transparent so that the eggs are then suspended in a mass of colourless jelly which adheres to the vegetation beneath the water.

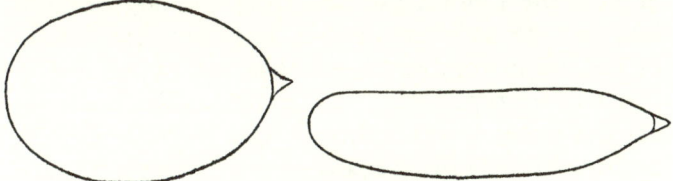

FIG. 2. Dragonfly eggs.

Again, others of the thick-bodied dragonflies and all the thin-bodied ones possess an apparatus at the hind end of the body which enables them to pierce the tissues of the water-plants and to insert an egg into each puncture, the insect, while carrying on this process, sitting upon a leaf of a plant floating on the water, submerging its long body and bending it beneath the leaf to insert the eggs into the tissues. The shape of the egg varies according to its destiny, the egg which is to hatch in the mud or attached to the vegetation being broadly oval and more or less the shape of a hen's egg, while the egg inserted within the tissues of the plant is elongated (*v.* Text-fig. 2).

Although, in a general way, the life history of the dragonfly was known, the details had not been worked out by anyone until I took up the subject about seventeen years ago. It was known that the egg was laid in the water, that a creature known as a 'nymph' hatched from the egg, that the nymph grew up in the water and ultimately crawled out, the perfect insect escaping through the breaking of the skin along the back of the nymph. It was not known how long the eggs took to incubate or how the nymph escaped from the egg or how long it took to grow up—in fact very few details of the life history had been observed by anyone.

I had become interested in dragonflies because, at that time, I was living in the Norfolk Broads district where a great many of the British kinds are to be found, and I even had the pleasure of discovering there a species new to Britain. I therefore determined to try and work out the life history and, as it was necessary to choose some particular species to work with, I selected one of the commonest of the thin-bodied kinds, a species of *Agrion*.

It was quite easy to obtain eggs by merely walking along the dykes or drains in the fens and watching the dragonflies ovipositing in the floating leaves of water-plants but, in gathering such leaves, one could not be sure that all the eggs in the tissues —which could easily be seen by holding up the thin leaf to the light—were the eggs of the particular kind of dragonfly which had been watched, and therefore, in order to make certain that I knew which species I was working with, I constructed a cage of muslin upon a wire frame, $15 \times 12 \times 12$ inches, and placed this over a large photographic developing dish half-filled with water. Upon the water I placed a few leaves plucked from the 'frog-bit' (*Hydrocharis morsus-ranæ*), having first made sure that they contained no dragonfly eggs, and I then captured six pairs of one kind of Agrionid dragonfly and placed them in the cage.

Now the dragonfly, like so many other insects, is a child of the sun and is only active in the presence of sunlight. On a dull day or if the cage was shaded, the insects sat about upon the muslin sides, scarcely moving hour after hour, but in full sunshine they flew about freely and ate readily any small flies and other insects which crawled upon the muslin. In order to supply them with food I swept a muslin net through the herbage along a hedgerow and thus captured numbers of small insects and these I released inside the cage, those which could not fly crawling about on the muslin sides. During the sunny periods the females settled upon the floating leaves and, submerging their long bodies, they bent them beneath the leaves and inserted their eggs, laying a number without changing their position so that the eggs, each in its own puncture, were arranged in arcs of circles.

Having obtained the eggs I released the captives and placed the leaves in a tumbler of water, covering the top of the tumbler with a piece of glass to keep out the dust. Now the leaves had been plucked from the plants so that within a fortnight they began to show signs of decay and in another week they had sunk to the bottom and become a mass of decayed vegetable material. Under natural conditions, of course, the leaves containing the eggs would have remained attached to the plant and would have continued to live, little the worse for the presence of the eggs, and I thought that I had made a fatal mistake and that all my work had been in vain. However, about a week later and almost four weeks after the eggs had been laid, actually on August 28th, a number of minute nymphs appeared upon the mass of decay in the tumbler—in fact most of the nymphs hatched out at the same time. In this way I missed, at that time, seeing the nymph hatch from the egg and it was not really until a year later that I actually saw how this happened, but as I want, in this lecture, to follow out the life history, I will describe it here, so that everything may come in the proper order.

The egg-shell of the dragonfly is transparent so that, once I had discovered how to keep the eggs alive and under observation after cutting them out of the plant tissues, it was possible to watch a great deal of the development of the embryo within the shell. I kept the eggs alive by laying them upon wet cotton-wool. If they were allowed to sink in the water—which seemed a most natural way to keep them—they died, and it seems therefore that they require a certain amount of air during the incubation period and, while wedged in the plant tissues, they obtain this from the air-spaces in the plant. This necessity is specially peculiar because of the fact that those dragonfly eggs which are laid in masses of jelly, or those which hatch in the mud at the bottom of a pond, get on quite well without any special means of aeration, so that these thin-bodied kinds and the thick-bodied ones which oviposit in the same manner are perhaps rather less well adapted to the aquatic life than the others.

I need not here go into details with regard to the development of the embryo and it will be sufficient to say that, when the nymph is ready to hatch, it occupies the shell from end to end, except for a small space in front of the head, the tail end being bent beneath the body. For some days before the time arrives for the nymph to escape, it can be seen making slight movements within the shell, but the first evidence that emergence is going to take place is the cracking of the shell just in front of the head beneath a small projection or pedicel which always exists at the front end of the egg. The crack extends round the shell and from it there slowly bulges out a very thin membrane which, as it extends, carries up the pedicel away from the rest of the shell, and this membrane, filled with fluid, continues to expand for an hour or

more. When it has attained its full size, which is
not very great in comparison with the size of the
egg, the head of the nymph begins to bulge into
it. Very slowly the head occupies the whole
vesicle and then, quite suddenly, the membrane
bursts and the nymph, wriggling violently, flows
out of the egg-shell.

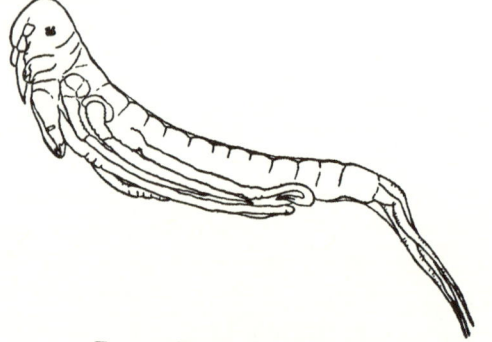

FIG. 3. Dragonfly pronymph.

When I first saw this happen, after having sat
with my eye at the microscope for more than two
hours, for the egg is only about $\frac{1}{20}$ inch long, I was
greatly disappointed because the creature which
emerged was a helpless cripple which lay upon its
side, unable to straighten its legs which remained
bent backwards beneath the body (*v.* Text-fig. 3).
In about two minutes, however, the cripple began
to wriggle and I saw the skin split open along the
back and there came out a neat little creature
which at once straightened its legs and arranged
its mouth parts and was obviously a perfect nymph.

There is, therefore, what I have called a 'pro-nymph' stage in the dragonfly or, in other words, the nymph which hatches from the egg moults in about two minutes after hatching. By observations later upon one of the thick-bodied dragon-flies (*Sympetrum*) I found that the same 'pro-nymph' stage occurred, but that it lasted rather longer, seven minutes being about the time of its duration.

Now these minute creatures which, in body length, were just about the length of the egg from which they had emerged, each carried at the posterior end three narrow pointed structures each of which was about as long as the nymph. In order to follow out the life history it was necessary to follow out the life histories of individual nymphs and, for this purpose, I obtained a number of tumblers. Each tumbler was then nearly filled with water and in each I placed a small piece of clean pond-weed to afford foothold for the occupant. One little nymph was then placed in each tumbler, a piece of glass being used as a cover for each little aquarium in order to keep out the dust. Each tumbler was labelled and every incident in the life of the occupant was noted.

The first question was how to feed these minute creatures and this was solved by collecting some dry horse-dung and placing it in a jar of water. In a few days the water in the jar was swarming with slipper animalcule (*Paramæcium*) and, by sucking the fluid up in a pipette and injecting a little into each tumbler, I was able to keep the little nymphs alive. As they grew larger the food was changed and consisted mainly of water-fleas and other minute organisms and also of small insect nymphs and larvæ. At first every nymph was examined and measured under the microscope every other day, but as there were more than seventy I only kept this going for about a fortnight, after that being satisfied with a bi-weekly examination.

As the weather became colder, the little creatures became very sluggish until, by November, they were showing but little interest in the food provided for them. Therefore in the beginning of December I placed eighteen of the tumblers in an incubator which was kept at a temperature of about 19° C. (66·2° F.), which is a rather high summer temperature for water. The effect of this warmth was very soon apparent as the occupants showed great activity and readiness to feed and they grew very rapidly so that, by the end of February, the average length of those in the warmth was almost $\frac{1}{2}$ inch (11·0 mm.) while that of those in the ordinary temperature of the laboratory was only about $\frac{1}{8}$ inch (2·7 mm.). All the nymphs which spent the winter in the incubator were full-grown within ten months of hatching from the eggs while those outside, with one single exception, were not full-grown until about a year later, so that the effect of the warmth in winter was very definite.

Now the nymph is specialized for its existence

in the water and there are certain points about it which are of interest. I have already mentioned that the dragonflies are divided into two groups, which, for simplicity, we have called the thick-bodied and thin-bodied kinds, and the nymphs of the two groups differ in certain important characters though they agree in others, and the two can be at once separated by the fact that whereas those of the thin-bodied group have three leaf-like projections at the tail end of the body, those of the thick-bodied group have not these lamellæ.

All the nymphs agree in certain characters which I will first describe. Under the head is a peculiar structure which is really a highly specialized lower lip and is usually known as 'the mask.' It is capable of being shot out in front of the mouth and, at its apex, bears a pair of claw-like structures with which it can seize anything coming within reach and, having done this, by withdrawing the mask beneath the head, the captured object is brought within reach of the powerful jaws which chew it up. Thus the nymph captures its prey by lying in wait for it or by slowly creeping about on the water-plants or on the bottom of the pond until its large eyes observe something moving within reach. It is curious that, although the eyes are very large, the sight appears to be very bad since the nymph only seems to see objects which are actually moving, and it cannot recognise food by the smell. For instance, if a piece of chopped worm is lying in front of the nymph it may so remain indefinitely without attracting attention but, if the food is made to move by means of a fine wire, it will be at once seized and devoured.

Another point in which all the nymphs agree is in the process of development. The outer skin is cast a number of times, each operation being described as a moult, the intervals between the moults being described as 'stages' in the life of the nymph.

Now almost all insects moult several times during the growing period and, whatever complications may be concerned in the phenomenon, one reason for casting the skin is easily given.

The insect skin is composed of a horny material called 'chitin' which, when new, is capable of being greatly stretched but which, as it gets old, becomes less and less expansible so that, as the insect continues to grow, this skin becomes too small for it. Under these circumstances the creature does an extraordinary thing. As soon as its coat begins to get tight, it commences to construct a new one inside the old one, and the new coat, when complete, cuts off the supply of nourishment to the old one. A line of weakness then develops along the back of the old coat in the front half of the body and the insect, working its tail end forwards inside the old coat, forces its back through a slit which appears along the line of weakness. Having humped its back through the gap, it next withdraws its head with the fine 'feelers' or antennæ and the mouth parts from the old skin and then the legs and, having got so far, it walks out of the old shell and the moult is complete. I have left out certain important points connected with the moult but I have said enough for our present purpose. Whereas some insects only moult three times during the growing period, others moult more often and the dragonfly nymph moults from ten to fifteen times, the number apparently depending partly, at least, upon what I might call the health of the nymph, that is, if the food supply is abundant, and conditions favourable, there will be fewer moults.

Another point in which all the nymphs agree is in the development of wing-pads on the back. The newly hatched nymph has no wing-pads and not until two moults have taken place is there any sign of these but, after that, at each moult they increase in size until, from being little bumps, they come to extend backwards over several segments of the body.

Now the nymph, although it lives in water, requires air for breathing purposes, and the air is carried in the body in a much branched system of tubes, there being two main air-tubes lying one along either side of the body. In the adult dragonfly and in all insects which live in the air, these tubes communicate with the outside air through a number of pores or 'spiracles' along the sides of the body. In a number of insects which live in the water some of these spiracles remain open and the insect consequently requires to come to the surface at intervals, bringing the spiracles into contact with the air so that the supply in the tubes may be replenished. But in some aquatic insects, including the dragonfly nymph, the spiracles are all closed and the necessary air is obtained from that dissolved in the water, the insects not requiring therefore to come to the surface at intervals. The fish breathes the air dissolved in the water in exactly the same way and one only has to exhaust the dissolved air and it will be found that neither fish nor dragonfly nymph can survive.

And now we come to one of the important differences between the nymph of the thin-bodied and that of the thick-bodied groups, as the two groups obtain their air-supply differently. The thin-bodied nymph has a comparatively thin skin and usually remains transparent for most of its life so that, if examined under the microscope, its heart can be seen beating and the boat-shaped blood corpuscles can be seen coursing through the body spaces, the elaborate system of air-tubes throughout the body being plainly visible. This type of nymph absorbs all the air it requires through its skin, and the three lamellæ at the tail end, in which the air-tubes form a branched system reminding one somewhat of the venation of a leaf, assist in this absorption of air and thus act as gills. These leaf-like outgrowths are not, however, essential to the life of the nymph and, if they are cut off, the insect survives in spite of their loss. It may be that, in their absence, growth is slower but that is a point which I have not yet investigated. Incidentally, however, if the lamellæ are removed they are slowly replaced, suggesting that they are of some importance in the economic life of the insect. But, when the nymph reaches its last stage before becoming a perfect dragonfly, a pair of spiracles appears upon it and then it frequently loses its 'tails,' whether accidentally or on purpose I don't know, and it can be seen travelling to the surface of the water at intervals and bringing one or other of these spiracles which are close behind the head, one on each side of the body, into touch with the air.

The lamellæ are also of use in another way. If the insect wishes to move rapidly from one place to another it wriggles its body rapidly from side to side and thus swims through the water, the 'tails' acting together as an organ of propulsion.

The nymph of the thick-bodied dragonfly, easily recognisable by the absence of 'tails,' similarly breathes the air dissolved in the water but it obtains it in a manner very different from that of the thin-bodied nymph. The skin of these larger forms is apparently too thick to allow of a free passage of air and therefore a special breathing apparatus has been developed in the hind end of the gut. Here there is a sac-like enlargement, the walls of which are closely beset with fine branches from the main air-tubes and the insect sucks water into this sac by expanding it, and the air-tubes take the air out of the water, which is then driven out again by the contraction of the walls of the sac. There is therefore a constant current of water in and out of this sac and, in order that the latter shall not become clogged by particles of matter being sucked in with the water, the entrance is protected by five sharp points which act as a filter and catch any solid object which tends to be drawn in with the current.

Just as the 'tails' of the thin-bodied type have a secondary function as organs of propulsion, so this 'sac' acts for the thick-bodied type, and the insect, on being disturbed, tucks its legs into its sides and by rapid intaking and expelling of the water drives itself forward, looking not unlike a small torpedo as it darts along.

Comparatively insignificant structures of the nymph are a pair of small feelers or 'antennæ' upon the head but, in watching the growth and development of the occupants of my tumblers, my attention was caught by the fact that on individuals of different sizes, the little structures were composed of different numbers of segments. I found, for instance, that the newly hatched nymphs had three segments only, whereas fully grown ones had seven and, in recording the details of the life of each individual, I noted the stages at which the increases took place. Now the antennæ slowly increase in length as the nymph grows and I was led to investigate where this increase in length occurred. It seemed probable that each segment would increase slightly at each moult, so that all would equally contribute to the increased length of the whole organ. It was easy to test this by measuring the proportionate length of each segment to the length of the antenna for each stage in the life of the nymph. In order to do this it was necessary to measure great numbers of antennæ of nymphs in different stages and thus obtain an average length for the antenna for each stage and the average length of each segment for that stage and, by making a chart of these average results, the fact appeared that the region of growth in the antenna lies in the third segment from the base— as it is the only segment which increases in length in proportion to the length of the antenna, during the life of the nymph. I have only mentioned this as an example of how a little curiosity may lead to an interesting result, and subsequent observation has shown that, in some other insects, the centre of growth of the antenna is similarly situated nearer the base than the apex, and this may perhaps be true for the antennæ of many others.

The emergence of the adult dragonfly takes place out of the water, the nymph crawling up some projecting support some time—half an hour or more—before the event is to take place. Having crawled up, the insect seems to go to sleep, holding on firmly to the support, and nothing further happens until the skin has completely dried. It is obvious, therefore, that a fine day is essential for the emergence and, as far as my experience goes, a bright morning is always selected for the operation. Slight wriggling movements of the body are the first indication that things are going to happen and a slit appears down the middle of the back involving the back part of the head and the part lying between the wing-pads. Through this slit a white or pinkish body bulges and the insect emerges by a normal moult from the nymph skin. But, having cleared its crumpled wings from the sheathing pads and having withdrawn its head and legs from their cases, it throws itself backwards so that it hangs suspended from the skin by its hind body which rapidly lengthens and, during this process, the colour changes and becomes darker. Once the body has assumed its correct length, the insect makes a sudden movement by which it once more straightens itself and seizes hold, either of the old skin or of the support above it, with its legs, and it then withdraws its tail from the nymph skin which it has now finished with. Its wings now quickly expand and assume their normal size, but it is not until an hour or more after this has occurred that the insect attempts its first flight, and not until several hours later that the wings become completely dry and the body assumes its full colouring.

Dragonflies go through their life history under very diverse conditions. While many prefer the still waters of the pond, others inhabit the river, and the struggle for existence is such that some kinds have been driven to take up their abode in strange situations. One or two are believed to come out of the water at night and seek their food on land; some frequent waterfalls where they must live a precarious existence. In tropical forests, areas of water are rare, in fact it has been said that tropical forests and permanent marshes are, to a large extent, mutually exclusive and, although this has meant the exclusion of large numbers of water-frequenting species, not a few have taken up their residence in small accumulations of water which occur in the vegetation, as, for instance, in holes in trees or in the broken stems of bamboos. In many tropical plants, such, for instance, as the banana or the pineapple, the leaf-bases so fit against the rest of the plant that water is held, either as a thin layer or even as a little pool, and in this water quite a number of insects of different kinds, and some other water-loving creatures, have taken up their abode. There are, for instance, the larvæ of certain gnats or mosquitoes and other flies; there are caddis larvæ and water-beetles both as larvæ and adults and there are other insects, including certain dragonfly nymphs. And this peculiar community is not confined to one part of one country or even to one continent but, in all parts of the tropics where this sort of habitat is available, water-insects have become adapted to it.

One of the most interesting of the dragonflies found in such situations is one whose habits were only discovered in 1908 when nymphs were found in Mexico, the life history being worked out later by Prof. Calvert, who found the insect in Costa Rica. This dragonfly, which goes by the name of

Mecistogaster modestus, is one of the thin-bodied group and is characterized as an adult by the great length of its body which is nearly 3 inches long, although the full-grown nymph from which this adult comes is only about ¾ inch in length. It has been suggested that the great length of the body of the adult is connected with its association with the pineapple plant, as without the long body it would be impossible for the female to reach down to the water between the leaves when she wanted to lay her eggs.

That the nymph lives and thrives in this situation is sufficient evidence that there is an abundant supply of food available, this food consisting of water-fleas and various small insect larvæ.

In the Sandwich Islands, where stagnant water is scarce, there are several kinds of thin-bodied dragonfly, the nymphs of some of which inhabit accumulations of water in the leaves of lilies, as was discovered many years ago, and here again we find that the adult dragonfly—in one case at least —possesses a very long body.

Judging by what we see of them in this country, we should scarcely call dragonflies common insects, and yet in some parts of the world they are extremely abundant, so abundant in fact that they are caught and used as food. Wallace, in his fascinating book on the Malay Archipelago (p. 119), tells us that in the Island of Lombok

"every day boys were to be seen walking along the roads and by the hedges and ditches, catching dragonflies with bird-lime. They carry a slender stick, with a few twigs at the end well anointed, so that the least touch captures the insect, whose wings are pulled off before it is consigned to a small basket. The dragonflies are so abundant at the time of the rice flowering that thousands are soon caught in this way. The bodies are fried in oil with onions and preserved shrimps or sometimes alone, and are considered a great delicacy."

Apart from this, it cannot be said that dragonflies are of great importance except that, in feeding upon other insects, they no doubt destroy many which are themselves harmful to mankind.

It is well known that certain dragonflies readily migrate and are frequently captured far out at sea. One common British species, the four-spotted Libellula (*Libellula* 4-*maculata*), is a well-known example, and large swarms have been recorded many times from various parts of Europe. Almost every year the Island of Heligoland is invaded by them from the continent and many stragglers reach these islands. In other parts of the world other kinds are known to form swarms and pass away but, however numerous these swarms may be, they never seem to increase permanently the number of dragonflies in any particular place. In parts of South America there occur what have been described as 'dragonfly storms' which are vividly described by the late W. H. Hudson in *The Naturalist in La Plata* (p. 130)—one of the most interesting books on Natural History that anyone could desire. In referring to these storms he says:

"The really wonderful thing about them...is that they appear only when flying before the south-west wind called *pampero*, the wind that blows from the interior of the pampas. The pampero is a dry cold wind, exceedingly violent. It bursts on the plains very suddenly and usually lasts only a short time, sometimes not more than 10 minutes; it comes irregularly and at all seasons of the year but is most frequent in the hot season and after exceptionally sultry weather. It is in summer and autumn that the large dragonflies appear: not *with* the wind but—and this is the most curious part of the matter—in advance of it; and inasmuch as these insects are not seen in the country at other times, and frequently appear in seasons of prolonged drought, when all the marshes and water courses for many hundreds of miles are dry, they must, of course, travel immense distances, flying before the wind at a speed of seventy or eighty miles per hour. On some occasions they appear almost simultaneously with the wind, going by like a flash and instantly disappearing from sight. You have scarcely time to see them before the wind strikes you. As a rule, however, they make their appearance from five to fifteen minutes before the wind strikes; and, when they are in great numbers, the air, to a height of 10 or 12 feet above the surface of the ground, is all at once seen to be full of them, rushing past with extraordinary velocity in a north-easterly direction. In very oppressive weather and when the swiftly advancing pampero brings no moving mountains of mingled cloud and dust, and is consequently not expected, the sudden apparition of the dragonfly is a most welcome one, for then an immediate burst of cold weather is confidently looked for.... Of the countless millions flying like thistledown before the great pampero wind, not one solitary traveller ever returns.... When they pass over the level treeless country, not one insect lags behind or permits the wind to overtake it, but, on arriving at a wood or large plantation, they swarm into it, as if seeking shelter from some swift pursuing enemy and, on such occasions, they sometimes remain clinging to the trees while the wind spends its force. This is particularly the case when the wind blows up at a late hour of the day; then, on the following morning, the dragonflies are seen clustering to the foliage in such numbers that many trees are covered with them, a large tree often appearing as if hung with curtains of some brown glistening material, too thick to show the green leaves beneath."

Dragonfly storms are also known farther south, in Patagonia, and Hudson mentions a case in which, at a race meeting near the town of El Carmen, a violent pampero came up, laden with dense dust clouds and, just before the storm broke, millions of dragonflies arrived and the insects, instead of rushing past in the usual manner, settled upon the men and horses so that these were quickly covered in clinging masses of them.

In working out the details of the life history of the dragonfly, I was merely satisfying my curiosity which, after all, is the real justification for doing any piece of scientific work, but in the experiment with the incubator by which I was enabled to bring on in one year nymphs which ordinarily take at least two years to mature, there is possibly an indication of the cause of the irregular superabundance of dragonflies which, for one reason or another, is the primary cause of the migratory movements of these insects. If, under normal climatic conditions, a dragonfly takes two or three years to pass through the nymph stage, exceptional conditions, such as concentration of food or, in colder climes, a particularly warm winter, might bring on a whole crop so that, instead of taking two or three years over the nymph stage, they might become full-grown in one. This would mean that when the time came for the emergence of the adults there would be two or three crops emerging at the same time—those which were in, say, their third and normally final year, those which had passed two years and would, under normal circumstances, have taken another, and those which had only done one year.

As some slight evidence in favour of this view I may mention the apparently irrelevant fact that there is "a widespread belief that the appearance of a swarm of dragonflies heralds a drought[1]." Tillyard recognises the truth of this which he explains by mentioning that, on the drying up of water holes in the Sydney district of Australia, he has noticed that, of all the larger aquatic life of the pond, the dragonfly nymphs last out the longest and he points out how, as the water decreases, the amount of food becomes concentrated and therefore more easily obtained. It may well be that the extra food supply has increased the rate of growth and forced on the development of nymphs which under other circumstances would not have emerged for at least another year.

You will see therefore that, in undertaking to work out a life history, such as that of the dragonfly, one opens up new problems in the course of one's observations. There are still all sorts of experiments to be made upon these dragonflies, such as the repeated removal of the 'tails' of the

[1] Tillyard, R. J. *The Biology of Dragonflies*, 1917, p. 334.

nymph to see how it will affect the rate of growth, and one has only to keep an insect under observation for a short time in order to begin to ask oneself all sorts of questions, and the asking of such questions may lead one to endeavour to discover the answers to them—and that is scientific research.

THE HABITS OF THE WATER-BEETLE

BEETLES form one of the higher groups of insects, those which, like the butterfly, pass through four stages in the life history—egg, larva, pupa and perfect insect, instead of three, like the dragonfly—egg, nymph and perfect insect. Beetles are characterized by having four wings, the upper pair of which, no longer of any use as organs of flight, have become hardened and lie along the back so as to form a protection to it and to the second pair of wings, which have become enlarged and, when not in use, are folded beneath the hard upper pair, and it is because of this condition of the upper pair of wings that the beetles have received the name Coleoptera, which literally means 'case-wings.'

We saw, in the dragonfly nymph, an insect very beautifully adapted to a life in the water in that it is capable of breathing the air dissolved in the water and can remain below the surface for practically the whole of the nymphal period. The dragonflies are what we can describe as an Aquatic Order of insects, since all the nymphs live in the water, but the beetles, on the other hand, we should describe as a Terrestrial Order because the vast majority of the different kinds spend the whole of their life on the land. At different periods, however, in the history of the beetles, attempts have been made by some of them to adapt themselves to a life in the water and thus we find several groups of 'water-beetles' which have not descended from some one water-beetle ancestor, but represent the results of attempts on the part of various independent ancestors to become aquatic. Thus some water-beetles show better adaptation than others in that in some, the larval or growing stage is capable of breathing the air dissolved in the water —as we saw that the dragonfly nymph does—but no adult beetle is capable of doing this and all have to come to the surface at intervals in order to renew their supply of air. Moreover, no water-beetle spends the whole of its life in the water, the larva, in almost every case, creeping on to land before it changes to a pupa, and in those cases where it does not leave the water to do this, it seals itself up under water in such a way that the cocoon which it makes is kept filled with air.

The adult water-beetle, although, unlike the adult dragonfly, it spends most of its life in the water and feeds there, leaves that element at times and, being possessed of wings, it flies to other pieces of water and, in this manner, is largely responsible for the dispersal of the species.

Although there are a number of widely separated groups of water-beetles descended from independent ancestors, there are two main groups, one of which is characterized by being a flesh-eating group in both adult and larval stages, while the other, although a flesh-eating group in the larval stage, is mainly a vegetarian group in the adult stage, and it is chiefly with the habits of representatives of these two groups that I intend to deal in this lecture. The representative of the flesh-eating water-beetles which I shall take is Dytiscus or Dyticus, the Great Diving beetle, a common inhabitant of our ponds and a beetle of reasonable size, being about an inch and more in length and perhaps $\frac{3}{4}$ inch in breadth. It is a handsome olive-green beetle and, when alive, its back gives one the impression of green mother-of-pearl, the effect disappearing when the insect dies.

There are marked differences between the male and the female. In the first place the male, which is usually rather the larger, has the front feet expanded into circular discs while the female has them as simple and narrow structures. In the second place, the upper wings or wing-cases are smooth in the male while in the female they are normally marked with longitudinal grooves. But here we notice an interesting thing. The aping of the male by the female is not confined to the human animal for we find that, in Dytiscus, the female occurs in two forms, one with fluted wing-cases and the other with smooth ones; it will be noticed, however, that she does not attempt to copy the large feet of the other sex.

We have six different kinds of Dytiscus in the British Islands but in only one of them do we find

PLATE VII

The 'tubs' and other aquaria for keeping water-insects and observing their habits.

The laboratory table showing the tumbler system for rearing small water-insects, and developing dishes for those kinds which will not attack one another.

the smooth form of female here. It is curious that in the case of this species this smooth form is the commoner one in eastern England, whereas farther west, for instance in Ireland, the fluted form is the common one, the smooth form being distinctly rare. On the continent we find that the smooth female is common in several species, which perhaps suggests that a more equable climate, such as there is in western Britain, is less suitable to the smooth form than a climate in which there are greater extremes of temperature. Or it is possible that it is not extremes but high summer temperatures which are connected with 'fluteness,' since it has been said that in cold ponds with little food even the male tends to appear with fluted wing-cases[1].

However, this question as to the cause of the females having two different forms still requires to be investigated and is by no means limited in its occurrence to the genus Dytiscus.

As with the dragonfly, the complete life history of Dytiscus had not been worked out in detail when I became interested in it some fourteen years ago and as, at that time, I had obtained a supply of living specimens of our rare northern form, the Lapland Dytiscus (*Dytiscus lapponicus*) from the islands of Skye and Eigg, I determined to work with that species (*v*. Pl. VIII, fig. 1).

I therefore obtained some empty barrels and,

[1] *Vide* Miall, L. C. *The Natural History of Aquatic Insects*, 1895, p. 59.

having sawn them across, I obtained two tubs from each. Having carefully cleaned them, I fitted frames upon them carrying wire-gauze lids so that I could, at any time, remove the lid and get at the contents of the tub (v. Pl. VII, fig. 1). Each tub was then filled to nearly one-third of its depth with earth in which a few roots of the common water-grass (*Glyceria fluitans*) were planted and, having stocked the tubs with numbers of pond-snails, insect larvæ, freshwater shrimps, water-fleas and other things suitable as food for the beetles, I released about a dozen specimens of *lapponicus* in one of the tubs. The time of year was October and, to my surprise, about a month later, there was no sign of any of the beetles, which had apparently either died or escaped. I therefore emptied the water out of the tub and found all the specimens deep in the mud at the bottom and apparently dead, but as they were in a good state of preservation, I thought I might as well mount them as specimens for my collection. However, upon taking them into the house and washing them clean I noticed that if I extended one of the legs, it very slowly closed in against the body and further examination showed that all the specimens were alive but in a state of suspended animation, were in fact asleep for the winter. I have already said that these beetles cannot utilize the air dissolved in the water but have to come to the surface at intervals to renew their supply and yet this Dytiscus, on the approach of winter, buries itself in the mud and, as I after-

PLATE VIII

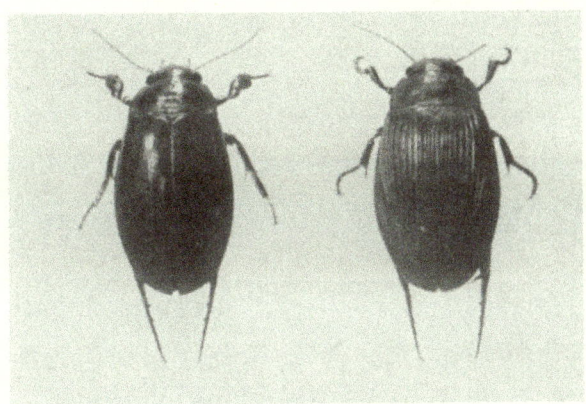

The Lapland Dytiscus (*Dytiscus lapponicus*) whose life history is outlined in Lecture V. (Life-size.) (Left, male: Right, female.)

A typical habitat of *Dytiscus lapponicus*. Lochan na Beinn Bhuidhe on the island of Eigg. In the distance the Scottish mainland is faintly visible.

wards discovered, remains there for about five months without breathing at all. Experiments in the laboratory showed that this winter-sleep depends upon the temperature since, in the warmed room, the beetles never buried themselves but remained active and fed all through the winter.

Towards the end of March and early in April, the beetles wakened from their winter sleep and were to be seen swimming about in the tub and feeding, and very soon after their awakening the females began to lay eggs. Now the female possesses a piercing apparatus, not unlike that of the female dragonfly and, by means of this, she bores holes in the tissues of the water-plants and in each hole she lays an egg, which is easily visible through the semi-transparent tissues.

I dissected out a number of these eggs and, after my experience with the dragonfly eggs, I laid them on wet cotton-wool where they mostly developed normally. As in the case of the dragonfly eggs, if they were submerged in water they always died, showing that air is necessary for the development of the embryo, air which, under ordinary circumstances, is obtained from the tissues of the water-plant.

Occasionally the eggs upon the wet cotton-wool died and I soon found that eggs which were not doing well became covered with minute creatures (Protozoa) moving about on the surface. Now with some minute water-animals, such as water-fleas, it has been found possible to stain some of the internal organs, which are visible through the transparent shell, without damaging the animal, and one of the materials which has been used for this vital staining is methylene-blue. I thought I should like to try staining some of the organs of the developing Dytiscus embryo, visible through the transparent egg-shell, and I discovered, quite accidentally, that if the cotton-wool upon which the eggs lay were wetted with a very weak solution of methylene-blue, the small protozoa never appeared upon the egg-shell and very few eggs died. By further experiment I found that if eggs upon which the protozoa had appeared were treated with methylene-blue, the protozoa were killed and the eggs usually survived.

The hatching of the Dytiscus larva from the egg is very different from that of the dragonfly nymph. Here, no vesicle expands from the front end of the egg into which the head later slowly moves. While I watched under the microscope a Dytiscus egg which was apparently ready to hatch, I was surprised to see the little creature within the shell moving its head up and down in a way which looked very much as if it were bowing to me. I noticed, however, that each time it made this movement, which it did at short intervals, two small prominences appeared, one on each side of the front of the head and that, at the summit of each of these prominences, was a minute spine. After discovering this, it was obvious that, every time the head moved up and down, the two minute spines, or one of them, came into contact with the shell and, after this had happened a number of times, the shell suddenly ripped over one of the spines and the larva, by actively wriggling, quickly emerged from the egg. On the cotton-wool the little creature was stranded but, on being lifted with a fine brush and placed in a tumbler of water, it sank slowly to the bottom, its legs spread out so that it landed comfortably upon its feet. And the same thing happened when the larva hatched from the eggs buried in the plant tissues. Once it had settled upon the bottom, the little larva rested for half an hour or more, after which it became restless and, using its feathered legs as oars, it swam to the surface where it brought its pointed tail up and, hanging head downwards with only its posterior extremity holding on to the surface film, it took in air at the only two spiracles which are functional. The little creature is transparent and the two main air-trunks with their branches are clearly visible through the skin. Both these air-trunks extend to the extreme end of the body where they open to the exterior by the two spiracles.

I adopted the same method with these larvæ as I had used with the dragonfly nymphs, that is, I used tumblers, one for each individual, placing a piece of pond-weed in the water to give foothold to the occupant, but, whereas the nymphs were very easy to keep alive, these Dytiscus larvæ appeared to be very delicate and a great many died (v. Pl. vii, fig. 2). It is possible to keep a number of dragonfly nymphs in one tumbler without fear of their injuring one another, so long as there is not a great difference between them in size, but Dytiscus larvæ have no respect for one another and four placed in a large tub were quickly reduced to one, so that, unless I was prepared to offer a tub to each larva, it looked as if I could not be successful in rearing this insect through to the pupa stage.

I came to the conclusion that the difficulty of keeping them alive in tumblers was partly due to the remains of the food which quickly fouled the water and, by changing the water in each tumbler every day, the death-rate was certainly lowered. I then tried the effect of adding methylene-blue to the water with the object of neutralizing any waste products and I found that this at once had a good effect. By experimenting with different strengths, I ultimately found that strong methylene-blue did not hurt the larva and that by supplying it liberally—so liberally that the water looked like ink—the larvæ lived better than in clean water, even when the contents of the tumbler were only renewed at intervals of several days. The only larvæ which I reared right through from egg to full size in the tumblers were some which spent almost the whole time in this deep-blue fluid.

As food for the smaller larvæ I used water-shrimps, the water-louse (*Asellus*) and small insect larvæ and later, as the larvæ grew, I relied almost entirely upon tadpoles which I collected in ponds and stored in the tubs, catching them as I required them for the tumblers.

The Dytiscus larva has a peculiar method of feeding. In front of its head are two large sharply-pointed formidable-looking jaws which can be spread wide apart to engulf the prey and which, the prey having been engulfed, can be brought close in front of the head. Each jaw has a fine tube running through it, one end opening just inside the sharp apex and the other opening on the upper side just near the base. When the jaws are closed the inner ends of these tubes communicate with the corners of the mouth, but when the jaws are open the tubes do not communicate with the mouth at all (*v*. Text-fig. 4). The mouth itself is peculiar; it is a long narrow slit across the face between the bases of the jaws, but if it is examined it is found that the upper and lower sides are grooved and ridged so that, when the mouth closes, the groove and ridge form a lock. Moreover, when the jaws are wide apart—ready to seize the prey—the mouth is open, but when the jaws are brought together in holding the prey, the mouth is closed so that, once the prey has been seized, the only communication into the mouth is through the fine tubes which, as has been said, open into the corners of the mouth.

Inside the mouth is a powerful sucking-pump which, by the action of the muscles, sucks the juices of the prey through the tubes in the jaws. But if this were the whole process of feeding, there would be a considerable waste, as the prey consists

largely of solid material and yet, if the remains
of the food are examined after the larva has cast
them aside, it will be found that very little remains
except the skin. For instance, of a large piece of
worm containing plenty of solid muscle, nothing
remains but a thin skin, if the larva has been
really hungry—and this is due to the fact that, at

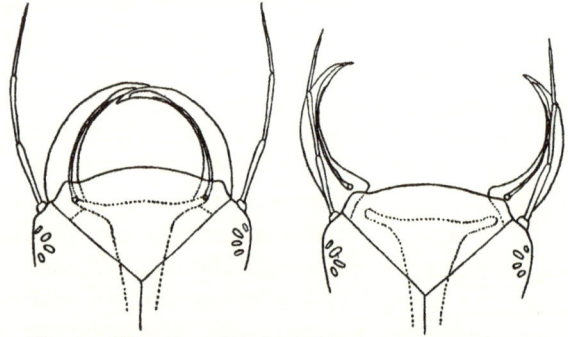

FIG. 4. Diagrammatic view of head of Dytiscus larva.

short intervals, the sucking-pump reverses its
action and pumps digestive fluid out through the
fine tubes in the jaws into the prey, where it dissolves away the solid material, which is then sucked
back into the mouth. At short intervals also, the
larva opens its jaws and gets a new grip of the prey
so as to work in a new spot and thus, little by little,
the solid material disappears as a fluid into the
Dytiscus larva.

This process is known as *external digestion*, the
digestion taking place largely outside the body
and only liquid food being taken in by the insect.

A great many insects feed by external digestion and various others, besides Dytiscus and its relations, have these tubes running through the jaws, so that we may say the idea has occurred to a number of different insects quite independently.

After about ten days, each larva in my tumblers cast its skin and again, about ten days later, it moulted, after which, being in its final larval stage, it continued to feed ravenously for four or five weeks, at the end of which time it ceased to take an interest in the food provided and spent its time swimming round just beneath the surface and evidently attempting to get out of the tumbler.

Now the full-grown larva leaves the water and burrows into the ground where it changes to a pupa, but I was anxious to see what it did beneath the soil. I therefore half-filled a glass-sided box with earth and, against the glass side, I made a burrow by pushing in a pencil more or less horizontally in a sloping bank of soil, and I was thus enabled to observe everything that happened in the burrow. By guiding the movements of the wandering larva to the entrance, I made it crawl into this, and as the burrow was only of a sufficient bore to enable the larva to crawl in, it was unable, once it had entered, to turn round. It therefore usually crawled to the inner end and then backed to escape, but I found that if I blocked the entrance with a stone, the larva, after shuffling backwards and forwards several times, usually took a philosophical view of the situation and made up its mind to adopt the burrow and, having done this, it commenced the construction of its cell at the inner end, by enlarging the space. The jaws, no longer necessary as feeding organs, are now used for moving pellets of soil from one place to another and to break up the pellets into fine particles, and the cell is mainly made by this process of breaking up and compacting, the head being used as a battering-ram in the first instance and the body, with the tail bent forwards over the head, being used later on to assist in this process. Very little earth is pushed into the unused part of the tunnel, but enough is so placed that the cell becomes completely cut off and, when finished, its walls are smooth and rounded.

The whole work occupies about twelve hours and the larva takes no rest during that time though, when the cell is completed, the labourer, utterly exhausted, lies in any position for about twenty-four hours, sleeping off its exhaustion. After waking up it takes up a definite attitude, lying more or less upon its back in a half-reclining position, and it remains thus, with only occasional movements suggestive of uneasy sleep, for from two to four weeks, after which the larval skin, splitting along the back, is cast off and a white pupa emerges.

On its back the pupa bears a number of short projecting spines and it has been suggested, in the case of another pupa with similar spines, that these were for the purpose of raising it off the damp floor of the cell. From what I saw, however, the pupa may rest in any position in the cell, very frequently on its face, so to speak, so that the use of the spines in this case is problematical—as indeed it is in the other case also.

The pupal stage lasts about three weeks and then the skin, splitting along the back in the usual way, sets free the perfect insect which is at first yellowish-white in colour and very soft. The colour gradually darkens and reaches normal after about forty-eight hours, but for a week or two the beetle remains immature, in that the skin takes a long time to acquire complete hardness. For a few weeks the beetle rests in the pupal cell and then, biting and pushing its way, it escapes from the earth and makes its way back to the water where, for some further time, it is easily recognised as one of the new generation by the comparative softness of the wing-cases.

A few words about the habits of the beetle itself will conclude this account of the life history which, in its main points, is the life history of all the relations of Dytiscus grouped together in the family Dytiscidæ.

Just as the larva requires to come to the surface at intervals to renew its supply of air, so does the adult beetle and, like the larva, it comes up tail first. But whereas the larva does this because the only openings into its air-tubes are through the two spiracles at its posterior extremity, the adult does it because it has, beneath the great wing-cases, a large air-reservoir, in which the hind wings lie and into which eight pairs of spiracles open, so that, although while hanging on to the surface film it may be renewing air in the tubes through the last pair of spiracles which are large, it is, at the same time, renewing the air in the reservoir.

Like the larva, the beetle itself is a ravenous feeder, and one has only to keep a specimen in an aquarium with other things to realize this. Not only will it devour other insects but it readily attacks its own brothers and sisters, and even minnows and sticklebacks are caught and devoured.

The beetles use their wings for flight but chiefly at night and, if kept in a tub without a gauze covering, they will depart if the conditions do not suit them. For a long time I kept an open tub filled with water in my garden in Cambridge, which I examined every three days, and quite a number of water-beetles used to appear in it, including occasionally a specimen of Dytiscus, but little seems to be known about the extent to which the beetles use their wings and it is possibly only at certain times of the year that the migratory instinct arises.

Having outlined the life history of Dytiscus, as a type of the flesh-eating water-beetle, I will take an example of the vegetarian group. There are two closely related kinds, the Great Silver beetle (*Hydrophilus*) and the Lesser Silver beetle (*Hydrŏus* or *Hydrocharis*), which differ a little from one

PLATE IX

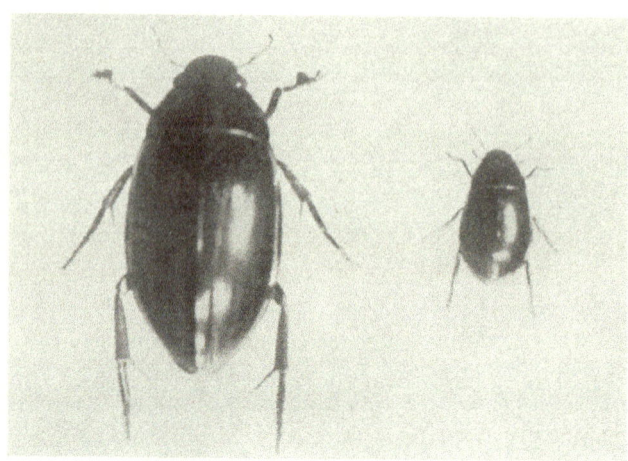

The Great Silver beetle (*Hydrophilus piceus*) and the Lesser Silver beetle (*Hydröus caraboides*). (Life-size.)

 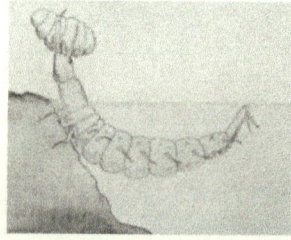

Drawings to show the different feeding-habits of the larvæ of the Hydrophilus and the Hydröus. The former feeds under water on pond-snails; the latter, having seized its prey under water, raises its head out of the water in order to eat it. In each case the tail end of the body, where the breathing pores are situated, is kept at the surface while feeding.

another in structure and habits, and I shall use the latter as my chief example but shall refer to the former in the course of the narrative (v. Pl. ix, fig. 1).

In the first place, the habits of the Hydrophilus beetles themselves differ considerably from those of Dytiscus. The Silver beetles receive their name from the fact that their undersides have a silvery appearance, owing to the whole surface being covered with a thin film of air, and this forms a reservoir in addition to that under the wing-cases which is similar to that in Dytiscus.

If the head of the Silver beetle is examined it will be found that, at the sides and towards the front, there are two club-shaped structures, the feelers or antennæ. In Dytiscus and its relations the antennæ are long tapering filaments, very different from those of the Silver beetles and their relations. For the most part these club-shaped antennæ are tucked away beneath the eyes when the beetles are in the water, but when the insects come out of the water the antennæ project freely in front of the head.

They are not, however, useless in the water because, if one of these beetles be watched, it will be seen to come to the surface at intervals, turn its head slightly sideways and push up one of the antennæ through the surface film. By this means it brings the film of air beneath its body into contact with the atmosphere, and as the film beneath the body is connected with the reservoir beneath the wing-cases, it renews its whole air supply through this one point of contact—a very different method from that employed by Dytiscus.

These beetles, although not incapable of swimming, are not such fine swimmers as Dytiscus, and mostly move about under water by crawling over the water-weeds or walking upside down upon the underside of the surface film.

The Lesser Silver beetle is almost entirely confined to the south and south-east of England and inhabits stagnant drains and ponds, and the same is true of the Greater species also. I began to experiment with the Lesser species when I was living in the north-east of Ireland where this insect does not naturally occur, but I obtained specimens from Surrey and kept them in my garden, a few miles outside Belfast, in tubs, each year obtaining eggs and rearing the larvæ and hatching out a new brood. It is quite clear therefore that the climate of north-eastern Ireland was quite suitable for this beetle and it is interesting for a moment to speculate as to why, then, this insect does not occur there. The conditions in my tubs were just such as were to be found in most of the ponds in Antrim and Down except that, in my tubs, the other forms of life were limited to what I chose to introduce and it seems, therefore, as if it is competition with other animals which limits the range of the species in the British Islands. It may be suggested, perhaps, that the reason the beetle does not occur in northern Ireland is because it hasn't been able to cross the Irish Sea, but

I am quite confident that I could have reared it equally well in southern Scotland or in any part of England in districts which no obvious barrier cuts off from its actual range.

Whereas Dytiscus lays its eggs singly in holes pierced in the plant tissues, the Silver beetles build elaborate cocoons which float upon the water and in which about fifty eggs are laid. The spinning of the cocoon is a wonderful process, as it is spun upon a film of air which comes from the under-side of the beetle and is added to from the supply under the wing-cases and this air is absolutely necessary, because the silk from the spinnerets will not form threads in the water. The female first finds a suitable place amongst the floating vegetation—the cocoon is always anchored to something —and then, holding on to the plant just below the surface, she turns so that her underside is upper-most and, slightly raising and depressing her wing-cases, she causes a bubble of air to expand over the end of her body where the spinnerets are concealed. Into this bubble she projects her spinnerets, a pair of fine flexible processes, and these are worked backwards and forwards upon the film of the bubble and a silken cocoon is thus formed. The two spinnerets, working side by side, each produce a fine thread and the two threads are spun parallel to one another but not so as to form a single thread. When the cocoon has only reached a stage of being a small cone, the female begins to lay her eggs, placing them carefully side by side and binding them together with silken threads spread over them to hold them in position. From time to time, as the cocoon increases in size, the mother raises and lowers her wing-cases, thus allowing more air to be added to the bubble, and of course she frequently brings one of her antennæ to the surface film in order to renew and add to her supply of air.

Egg-laying and cocoon-building are thus alternating processes and, the full complement of eggs having been laid, the cocoon is closed, about an hour after the commencement of the work. There is, however, always a space within the cocoon after all the eggs have been laid, this space being occupied by very loosely woven silk thread, and the cocoon terminates in an upwardly pointing process, sometimes called 'the mast,' which projects above the water. It has been stated that, if the mast is cut off from the cocoon, the eggs perish, but this is not true and it is rather curious that these eggs, unlike those of the thin-bodied dragon-fly and of Dytiscus, develop quite well if removed from the cocoon and submerged in water. I found however that they developed normally upon wet cotton-wool, so that it was possible to watch the whole development of the embryo through the transparent shell.

The egg hatches in from twelve to twenty days, depending, apparently, chiefly upon temperature, and the young larvæ spend a day or two inside the cocoon before biting their way out and it is, perhaps, partly to give them extra play room that space is left within the cocoon. When they emerge they are buoyant and float in the water and, if they are forcibly submerged, they float up again to the surface. If, however, the larvæ hatch from sub-merged eggs they are not buoyant and make great efforts, by crawling up the vegetation, to reach the surface. If they are successful they at once raise their heads above water and can be seen taking a 'drink' of air, which can be seen in their trans-parent bodies, passing down in the form of two or three bubbles into the gut. After that they are buoyant and this suggests that the supply of air in the cocoon may be to provide the larvæ, on hatching, with the necessary air drink or, supposing the cocoon has become waterlogged, the 'mast' would provide a ready way to the surface.

Now the larvæ of the Great and Lesser Silver beetles differ somewhat in form, in that the latter have lateral processes projecting outwards, these being absent in the former and, in the absence of these processes, the larger species differs from many others of its relations. Again, the larvæ of the Great Silver beetle differ in habits from most of their relations since they feed under water upon pond-snails only, whereas the relations for the most part feed above water, as we shall see.

I kept a number of the growing larvæ in tumblers so as to record the activities of each individual, and I also kept small numbers together in large shallow photographic dishes as, although they only eat animal food, they are not so voracious as Dytiscus larvæ and will not attack one another, unless there is great disparity in size or unless they are very short of other food.

Though the larvæ of the Greater species will only eat pond-snails—so far as my experience goes—those of the Lesser species will eat almost any kind of animal material and I fed them mostly upon chopped up earthworms. All these Hydrophilid larvæ, like the Dytiscid, have to come to the surface to breathe, and not being such excellent swimmers as the Dytiscids, they are seldom very far from the surface. The larva of the Greater Silver species, having found a pond-snail, holds it in its jaws and, bending its head backwards, it hollows its back in such a way as to hold the shell of the snail (v. Pl.. ix, fig. 2). The two jaws are differently constructed; the left jaw, on its inner face, has a small tooth not far behind the apex, while the right one has a double tooth on its inner face, not far from the base. When the snail has been firmly wedged into the hollow in the back, the larva, using the right jaw as a shell-opener, the double tooth doing the cutting, slowly turns the shell by movements of the left jaw and thus the shell is cut away and the occupant exposed and quickly sucked down into the mouth. The larva of the Lesser Silver beetle and other members of the family feeds differently from that of the Great Silver beetle. Having seized its prey, be it a water-flea, insect larva, tadpole or small worm—according to the opportunity and size of the predator—it moves rapidly about searching for a spot at which it can raise its head above water and, having found this, it partly crawls out, exposing the head and one or more body segments and, holding the food in its mouth parts, it works its jaws in and out so as to squeeze out the juices, at the same time regurgitating a digestive fluid which dissolves away the solid parts of the prey and even causes them to froth, the fluids being then sucked back into the mouth (v. Pl. ix, fig. 3). We see therefore that, as in the case of the Dytiscus larva, external digestion is carried out but there are no tubes through the jaws nor is there a mouth lock, so that the process is simpler here.

The larva casts its skin twice during its life, which occupies about six weeks, and at the end of that time it ceases to feed, swims about restlessly for a little and then leaves the water, burrowing into the soil like the Dytiscus larva and forming a very similar cell where, in the course of three or four weeks, it changes to a pupa. The pupal stage lasts another three or four weeks, after which the beetle emerges from the pupal skin, at first white, but gradually darkening through reds and browns to the blue-black of the mature insect. The whole life cycle from egg to perfect insect occupies from nine to thirteen weeks, but it is some time after the beetle has emerged in the cell that it breaks its way out and returns to the water.

We have now seen that two types of water-beetle,

the Dytiscid and the Hydrophilid, although in certain general points agreeing in their life histories, differ profoundly from one another when we come to enquire into details and that, as in the case of the dragonflies, differences in structure and habit may occur in closely related forms, differences which frequently appear to be so trivial and yet which mean a great deal to the insects themselves in the way that they fit into the communities to which they belong.

From the fact that it has been possible to rear dragonflies and these two types of water-beetles in tumblers, it must not be assumed that the habits of all water-insects or even all water-beetles can be observed in the same manner and, as a matter of fact, very different methods are necessary in many cases. For instance, there is a group of water-beetles belonging to a family far removed from either of the two already mentioned and these beetles go by the name of Donacia, or, as they are sometimes called, the Golden Water-beetles, because many of them exhibit bright metallic colourings. Most of the Golden beetle family live entirely upon the land, but a small group of them has taken to a life in the water and they have become adapted to that life in a manner very different from that seen in the types already described. In the first place the adults mostly spend their lives out of the water on the vegetation round the margins or projecting from the surface of ponds, lakes or streams, but in a few cases they spend almost all their time beneath the water, crawling about on submerged vegetation and enveloped in a thin film of air. These beetles lay their eggs upon the underside of leaves floating upon the water and the larvæ which hatch from them are as unlike the water-beetle larvæ already described as they can well be. They are dirty white in colour, tapering rapidly towards head and tail, the body being more or less cylindrical, and they possess three pairs of short legs and, when full-grown, are not more than half an inch long. They are very slow in their movements and suggest other beetle larvæ which live in the soil on dry land, and their chief peculiarity is a pair of sharp processes projecting posteriorly on the upper side of a segment at the posterior end of the body. Examination of these processes shows that they are hollow and have an opening near the apex and that the hollow space in the process communicates directly with the ends of the two great air-tubes which run forward, one on either side of the body. These larvæ, on hatching from the eggs, crawl down the plants to the roots which are in the mud and there they begin to feed, but they first drive the two processes I have mentioned into the tissues of the plant. By so doing, they tap the air supply which runs all through the plant tissues and thus they are enabled to breathe even though they are buried in the mud below the water. When full-grown, these larvæ surround themselves with a yellow oval cocoon which they secrete from the skin and which, in some way, they fill with air from the tissues of the plant and in this cocoon they change into pupæ. The pupal stage occurs in the autumn, but it does not last long, though the adult beetle remains in the cocoon until the following spring, when it bites its way out and crawls to the surface.

Such a life history cannot be worked out in tumblers, nor have I had any success with it in my tubs, but it can be discovered by observations in the field, by watching month by month or, if necessary, week by week or day by day, a particular locality and collecting facts and drawing conclusions. By examining the undersides of many floating leaves of water-plants in June and July, the eggs can be found in rows or circles; by pulling up water-plants and carefully rinsing the mud from the roots, the larvæ can be found in different stages of growth practically all through the year, while the cocoons can be found in the same situations from about August or September until May or June and, by opening numbers of cocoons during this period, the duration of the pupal period can be discovered.

I have described this life history of the Golden water-beetle as if it had all been worked out in a year although, as a matter of fact, it was only after a number of years of casual observations that it was completed, but each observation was recorded with date and full details so that, after a time, by looking up all the observations and gathering them together, it was possible to outline the life history.

Thus you see how it is possible to do a piece of rough work such as this without any great demand upon time. It is not a piece of work such as that upon the other water-beetles or the dragonfly—which was taken much more seriously and where much more detail was observed, but it will indicate to you that a great deal of amusement and interest can be obtained casually and that much may be learnt by methodical recording of casual observations.

Lecture VI

THE HABITS OF INSECTS AND
THE WORK OF MAN

IT may well be asked by some of those who have listened to the earlier lectures of this course, What is the use of all the work which has been described? And I can answer without any feeling of shame that, so far as I know, the work is of no use whatever and that my only object in undertaking it was that I was curious and wanted to know how the insects lived.

I believe that there is still an ancient Physical or Mathematical Society which, at its annual banquet, drinks the following toast: 'Here's to the Society; let no one ever do anything that is of any use to anybody.' This sounds a very selfish, unchristian wish but it is the true scientific attitude —to do the work for the work's sake and never mind whether or not it is going to be useful.

But a large amount of work done for its own sake turns out to be of use and, although we seldom recognise what we owe to the humble naturalist, he has contributed largely towards our health and comfort.

It is not so very long since it was first recognised what an important part insects play in the life of man and, although a vast number of kinds live their lives without in any way affecting man's affairs, the others can be described either as useful or injurious insects.

With regard to the useful insects, many kinds, but especially the bees, are useful in the pollination of flowers and, without them, many flowers would set no seed. When red clover seed was first taken to New Zealand a magnificent crop was harvested in the first year but, to everyone's surprise, no seed was produced. The simple entomologist explained that clover flowers were entirely dependent upon humble-bees for their pollination and that there were none in New Zealand. Steps were at once taken to introduce some from England and two kinds were sent out. The entomologist was, apparently, not consulted, otherwise some advice might have been given as to the best kinds of humble-bees to introduce and, in consequence, one of the kinds sent out was the so-called Terrestrial Humble-bee (*Bombus terrestris*) which is in the habit of getting at the nectaries of flowers by biting a hole in the side instead of entering by the tube, the latter being essential in order that the stamens with their pollen and also the sticky stigma may be brushed over. This bee, therefore, was not only less useful in New Zealand than another kind would have been, but it has done some damage in other directions, by stealing the nectar from other flowers in the same way, so that these are now setting less seed than previously.

Although honey is no longer of the importance

that it used to be when our staple drink was mead, yet it is a nutritive and pleasant food and we owe it to the hive-bee, which has been so long cultivated by man that it now ranks almost as a domestic animal. Silk is another useful commodity which we owe to the caterpillars of certain moths, the true 'silk worm' again being a domestic animal to such an extent that it does not occur wild in nature.

Shellac and cochineal come from two small bugs, the latter having become much less important since the discovery of the aniline dyes. Ink is manufactured with the assistance of tannic acid obtained from certain galls, which are special tissue growths of plants caused by the secretions of minute relations of the wasps.

In another way many insects are useful to man in that they prey upon other insects which are, in one way or another, injurious to man. The majority of these preying insects are what we call 'parasites' whose larvæ feed upon or within other insects, although this is not strictly the correct interpretation of the term 'parasite.'

In many important ways, therefore, insects are useful to man and, just as in the past he has cultivated those insects which supplied him with useful commodities, now he is cultivating and exploiting the parasites of injurious insects, as a method of waging war upon the latter. With regard to these injurious insects, some of them attack man and his domestic animals and, by their presence, cause diseases; flies are the chief aggressors in this case. Others transport the germs of disease and thus are indirectly injurious. For instance, malaria, sleeping sickness and typhus fever are examples of diseases the germs of which are transmitted from one person to another by insects. Others again attack our crops, our orchards and our forests and innumerable manufactured and preserved or stored articles, such as furs and feathers, leather and woollen goods, stored and preserved fruits, grain, flour, and in fact almost every article of commercial value.

It will give you some idea of the seriousness of this question of damage done by insects when I tell you that, until recently, half the human death-rate each year was due to malaria, transmitted from person to person by mosquitoes. In some parts of the world more than half the apple crop is damaged by the attacks of a small caterpillar which eats the flesh of the fruit—the caterpillar of the Codlin moth. In the southern United States the Cotton Boll weevil causes damage every year to the cotton crop to the amount of about £40,000,000[1] and it has been prophesied that this beetle may, in time, ruin the cotton-growing industry of the United States. In our own islands, insects levy a toll of about 10 per cent. of the value of our crops.

The vast importance of insects in the affairs of man has only come to be recognised comparatively

[1] Hunter, W. D. and Coad, B. R. 'The Boll weevil Problem.' *U.S. Dep. Agric. Farmers' Bulletin*, No. 1329. June, 1923.

recently and it may be said that entomology has only been recognised as of some use within the last fifty years.

My object in this lecture is to show how, in dealing with injurious insects, an accurate knowledge of the habits is necessary, not only of the injurious kinds but of all other kinds which may in any way affect the well-being of the injurious kinds—in fact, a study of the community to which the injurious insect belongs is closely associated with the study of the insect itself, and once this work has been done, the question of how to apply the acquired knowledge is the problem for the economic entomologist. In order to attain my object I propose to take a few examples and the first we might entitle 'The Mosquito and Malaria.'

The relationship of the gnat or mosquito to the transmission of disease was suspected as long ago as 1878 and, in 1879, Sir Patrick Manson first proved that the insect was implicated in the transmission of a tropical disease due to the presence of a minute worm, which undergoes part of its development in man and part in the mosquito. In 1894 Manson, arguing from this, suggested that the mosquito might be concerned in the transmission of malaria and, in the years following, Sir Ronald Ross worked out the whole life history of the organism which causes malaria. In 1898 it was proved by the experiments of two young London doctors that living in a malaria-infested locality did not lead to an attack of malaria, so long as precautions were taken to prevent the residents from being bitten by mosquitoes.

In 1881 an attempt had been made to show that yellow fever, like malaria, was only transmissible from one person to another through the agency of mosquitoes and this was definitely accepted, and proved as far as proof was possible, by the United States Commission on Yellow Fever in 1900. The life history of the malaria parasite, as discovered by Sir Ronald Ross, is of great interest, but I need only say about it that the minute organism undergoes certain changes in the blood of the human being, where it can continue to live for an unknown period, certainly many years, but in order to complete its life history, it is essential that it should, when at a certain stage, be sucked up in the blood by a mosquito, in which insect it undergoes further changes which occupy from about nine to perhaps twenty-two days[1]. The mosquito, after that period, feeding upon some other human being, pumps some of its saliva into the blood and with it some of the malaria germs, and thus and thus only does the second human being become infected.

Once it was clear that malaria and yellow fever and certain other mosquito-borne diseases were neither infectious nor contagious, it became clear that old methods had to give place to new and the question arose as to whether such diseases could

[1] *Vide* Prof. Warrington Yorke, 'The Malaria Treatment of General Paralysis.' *Nature*, Oct. 25th, 1924, p. 616.

not be stamped out by the destruction of the germ-carrying mosquito and, in order to test this, the life history of the insect had to be known to the last detail.

Now the mosquitoes and gnats are members of a family of the flies and are what we know as aquatic insects. The eggs are laid upon the water, by some kinds in 'rafts,' the eggs all being glued together side by side in an upright position, by other kinds, each egg floating like a minute boat upon the surface. The eggs are peculiar in that they are unsinkable and unwettable, being covered with a water-proof varnish, and the egg-rafts are self-righting and cannot be upset and also, in the case of the eggs laid singly, they are like self-righting life-boats. The objects of these peculiarities are, first, that the developing embryo may have a good supply of air—as it dies if the egg is forcibly kept under water—and secondly, that the larva, when it escapes from the egg, may escape beneath the water and not get caught in the surface-film, and as the larval head is in a definite position within the egg and must break the shell at a certain point, it is essential that that point shall be beneath the surface-film.

The larva, like that of the water-beetles Dytiscus and Hydrophilus and their relations, has to come to the surface at intervals to renew its supply of air, which it does by bringing to the surface-film the only pair of open spiracles which it has in connection with its air-tube system, and these spiracles are carried upon the upper side of the eighth body segment, in many cases at the end of a tube known as 'the siphon.' If the larvæ are prevented from coming to the surface, they drown. They feed upon minute organisms which swim in the water, which they catch by creating a current of water round the mouth by means of a pair of brush-like feelers or 'antennæ.' Other hairs on the organs surrounding the mouth catch the organisms as they are whirled inwards and they are then swallowed.

The larva is heavier than water, that is, it sinks unless, by violent wriggling movements, it propels itself upwards, but once it gets its spiracles into contact with the surface-film it can remain suspended there as long as it chooses, by merely spreading out on the film five little flaps surrounding the openings. If it wishes to sink, it merely contracts these flaps and lets go its hold on the film.

The convenience of being heavier than the water is obvious because the food is beneath the surface and the larva can thus remain in the midst of its food without having to make perpetual movements to keep down and, seeing that it has no legs, it has nothing by means of which it could hold on to anything to keep itself down. It is obvious also that having its breathing pores at the opposite end of the body from the mouth is convenient, since the larva can feed, hanging head downwards, while it is attached to the surface-film for breathing.

The larva is full-grown in from one to three weeks, according to temperature, having passed through three stages, determined by moults. At the completion of the third stage, the moult transforms the larva into a pupa, not only very different in form but different in certain important characters. In the first place, it is no longer heavier than water but floats to the surface and remains there unless it actively submerges itself. In the second place, instead of having the front end heavier than the tail end, conditions are reversed and the tail end is depressed, and thirdly, although there is still only one pair of open spiracles, these are now situated upon the back just behind the head. It is easy to understand why these changes have taken place. In the first place, the perfect insect will emerge in a few days from the pupal skin and the perfect insect inhabits the air, not the water, and therefore requires to escape into the air, and, as the moult which frees it takes place in the usual way, by the splitting of the pupal skin along the middle line of the front part of the back, it is necessary that this region should be in contact with the surface. Hence also the transposition of the breathing openings to this region. As it is necessary that the pupa should be quiescent at the time of the emergence of the mosquito, it is essential that it should float to the surface automatically and remain there without having to exert itself. Hence the change from the heavy larva to the light pupa.

These general facts of the life history, true of all the mosquitoes and gnats, were known long before the insects became notorious, the observations having been made by humble naturalists who merely wanted to know how these curious creatures lived. But, with the discovery of the connection between these insects and various dread diseases, the facts at once became important in connection with planning the destruction of the insects, and more facts were wanted as to the details of the habits of each different kind of mosquito and gnat, and it was on the knowledge of the habits of these insects that plans were laid for combating, with immense success, yellow fever and malaria.

As to how this success was achieved can perhaps best be demonstrated by telling of the campaigns undertaken against these diseases in the West Indies and Central America, first in connection with the stamping out of yellow fever in Havana, the chief town of Cuba, and secondly in connection with the building of the Panama Canal.

Havana had long been a hot-bed of yellow fever and every year had a very high death-rate. It is true that improved conditions had seen a reduction of the death-rate, so that in the years 1891–1900 the average was only about four hundred a year as against one thousand a year in the 'seventies. Yellow fever is transmitted from person to person by one kind of mosquito only, usually known as the Brindled gnat (*Aedes calopus*, or under the old name *Stegomyia fasciata*). One of the peculiarities of its habits is that it is practically a domestic species, very rarely found breeding far from the haunts of man and almost invariably in artificial water-containers such as water-butts, vessels of water standing in houses, old tins and such like.

Experiments in its control began in Havana in 1900 and in February, 1901, a vigorous campaign was started:

(1) An order was issued throughout the town that all receptacles containing water were to be kept mosquito-proof, so as to prevent eggs being laid in them.

(2) The town was divided into districts under sanitary inspectors and

(α) A house-to-house inspection was organized and receptacles found to contain eggs or larvæ were destroyed and the owners fined.

(β) Rain gutters were examined to see that the water did not stand in them and, if it did, the owner had to have them put right and kept in good condition.

(γ) No empty tins or other receptacles for water were allowed to lie about and all puddles were either filled up or treated with oil, the oil spreading over the surface and forming a film which prevented the larvæ and pupæ from renewing their air supply.

(3) Hospitals and houses containing yellow fever patients were screened with fine wire gauze through which the mosquitoes could not pass and thus, as they could not suck the blood of the

patients, they could not become infected with the disease and therefore could not transmit it to other people.

(4) Buildings, where yellow fever had occurred, and the adjacent houses were fumigated so as to destroy any mosquitoes which might have fed upon the patients.

There had been twenty-four cases of yellow fever in January and eight in February, the month in which the control measures were commenced. In March only two cases occurred and another two were recorded in April while four more broke out in May. In June there were none and it seems probable that there would have been no more but for the fact that someone suffering from yellow fever came in from an uncontrolled area in July and thus caused four cases that month and a further six in August. Only one case was registered in September and then no more from October to January—which had previously always been *the* yellow fever period. The man responsible for this successful attack upon the Brindled mosquito was Major, later Colonel, Gorgas. He showed that a knowledge of the habits of the insect could be turned to a useful purpose and he changed Havana from an unhealthy town into a healthy one.

It was just about this time that the question of building a canal across Central America arose, by no means for the first time. The first time this matter was discussed was about 1550 and after

that in the eighteenth and early part of the nine-
teenth centuries, various routes for the canal were
suggested. In 1855 a railway was built across
Panama to deal with the transport problem which
had arisen mainly through the discovery of gold
in California. This was very expensive, not only
in money but in lives, and it served at the same time
to again postpone the question of the canal. Then
the discovery of oil in California led to the construction of an 8-inch pipe alongside the railway,
through which oil was pumped from ships lying
at Panama on the Pacific side to ships at Colon,
on the Atlantic side. As recently as 1918 a similar
pipe was completed across Scotland, from Old
Kilpatrick, Glasgow, to the Grangemouth docks
on the Firth of Forth, following the course of the
Forth and Clyde Canal and 36 miles long—
about the same length as the Panama pipe—for
supplying our warships in the North Sea and
lessening the risk to the oil carriers from submarine
attack.

Between 1870 and 1875, the United States
Government surveyed a large part of Central
America with a view to deciding which of the six
previously suggested routes was the best for the
canal, and they reduced these six to two, one of
which was across Nicaragua and the other across
Panama, and it is interesting to note that a United
States Commission in 1876 reported that, of these
two routes, the former possessed greater advan-
tages and offered fewer difficulties.

However, in 1879, an International Congress met in France and ultimately, in 1880, a company was formed to construct the Panama Canal, and the work was carried on for a number of years under the management of M. Ferdinand de Lesseps. The company failed and the circumstances of the failure, which was partly due to extravagance, gave rise to all the talk about 'the Panama Scandal.' Another company was formed and that again failed and, in thirteen years, less than half, and that the easier half, of the canal had been completed at an expenditure of $260,000,000 and with an immense loss of life, more than 22,000 labourers having died between 1881 and 1889, the death-rate having amounted to 240 per 1000 a year. The death-rate was due mainly to malaria and yellow fever and thus the failure of the canal companies was largely due to the mosquito.

In spite of these failures the idea of the canal was not given up and it became more and more necessary from the point of view of trade, but in the end it was war which decided the matter. In 1900 the Spanish-American War began with the blowing up of the United States warship *Maine* in Havana harbour, and this ship was a very serious loss to the U.S. navy. There was nothing to replace it on the eastern side of America and consequently the battleship *Oregon* was ordered to the seat of war from the western side. This necessitated a voyage of 8000 miles round Cape Horn in the face of the Spanish fleet and, although it was successfully accomplished, the United States Government and nation unanimously decided that a short route must be established.

The wonderful work of Colonel Gorgas in Havana, already referred to, had just been completed and in other parts of the world the control of malaria had been carried out with some success. Another circumstance which made the moment very favourable was a revolution in Central America, the Government of Panama having broken away from the Colombian Federation and placed itself under the protection of the United States. The Government of the latter, therefore, made a treaty with the Republic of Panama by which a strip of land ten miles wide, stretching from the Atlantic to the Pacific, was leased in perpetuity to the United States, the strip being known as the 'Canal Zone,' and along the middle of this the Canal now runs.

In 1904 Colonel Gorgas was sent to take charge of the sanitation of the zone and he commenced by making war upon the mosquito. He first put the town of Panama into order by piping the whole water-supply, thus eliminating open water channels and the necessity for water-carrying and storing utensils. The city of Colon, at the opposite end of the zone, was similarly taken in hand and in two years both these centres were clear of yellow fever.

All sorts of experimental work to elucidate the habits of the mosquito were necessary in dealing with the malaria problem: how many kinds of mosquito carried the malaria parasite: how far mosquitoes would fly and how they became dispersed and so on. For this purpose mosquitoes were bred out and, in order to ascertain facts as to flight and dispersal, they were sprayed with coloured fluids and these marked individuals were liberated and records kept as to their places of capture. It was found that mosquitoes readily settle upon dark clothing and by this means may be carried long distances in trains and may thus also enter screened houses. The colouring of mosquitoes, and various other methods, proved the movements back and forwards over considerable distances between breeding and feeding grounds and the experiments greatly assisted Colonel Gorgas in arranging his methods of control. The breeding-places of the mosquitoes were sought for and, where possible, were filled in; where this was not possible the water was oiled, the film of oil spreading upon the surface causing larvæ and pupæ beneath to be suffocated, as they could not reach the air through the film. In other cases the water was treated with a fluid which mixed with the water and killed the larvæ; certain kinds of fishes, especially one from Barbadoes known as 'millions,' were introduced into stagnant ponds where they fed upon the larvæ and pupæ. Other measures adopted were the careful choice of localities for housing the labourers and the keeping apart of the dwellings of native and white workmen, since the former are almost always carriers of malaria; the screening of houses and verandahs and of hospitals and the insistence upon all white people sleeping under mosquito-nets.

It is quite safe to say that, without the work of Colonel Gorgas, the Canal either could not have been constructed or its cost in lives and money would have been many times greater, so that here again we see how a knowledge of the habits of insects has been applied for the benefit of mankind.

We therefore wage war upon the mosquito because, although itself nothing more than a nuisance to us, it transmits the germs of disease from person to person; but now I will refer to an insect which, of itself, affects our cattle and which, by damaging the hides, causes a loss in Britain alone of from £2,000,000 to £7,000,000 annually.

This is a rather handsome fly known as the Bot- or Warble-fly, whose larvæ live under the skin of horses and cattle and, in order to breathe, pierce the hide and expose their spiracles at the opening, the spiracles being placed side by side at the posterior end of the body. Just as we suffocate the mosquito larvæ by covering the water surface with oil, so we can suffocate the warble larvæ by rubbing grease into the warbles or 'worm-holes' and thus closing the spiracles of the maggot.

In olden times before the habits of these insects had been studied, it was considered possible to charm them out of the animal and, in the seventeenth century, Scot's *Discovery of Witchcraft* gives the following instructions:

You must both say and do this upon the diseased horse three days together before the sunrising: 'In the name of God the Father, the Sonne and the Holy Ghost, I conjure thee, O worm, by God the Father, the Sonne and the Holy Ghost, that thou neither eate nor drink the flesh, blood or bones of this horse; and that thou hereby maiest be made as patient as Job and as good as S. John Baptist when he baptized Christ in Jordan. In the name of the Father, Sonne and Holy Ghost,' and then say Paternosters and three Aves in the right ear of the horse to the Glory of the Holy Trinity.

In more modern times, although certain rites were observed, a practical knowledge of the habits of the insect entered into the ceremony. For instance, Miss Ormerod (1886) tells how, in Ireland, charms were being used against these larvæ:

The charmer is generally an old woman. When she enters the stable of the sick cow, she calls for some butter or lard. After this has been placed before her she prays for a time to some spirit. After the spirit of destruction is exorcized, she takes the butter and quickly covers the breathing aperture of the maggot and crosses it. The result of all this is that the maggots die and fall, or are easily picked out without causing any pain.

With regard to the insect pests of forests and crops, observation of habits shows in which stage a particular insect is most easily attacked and it further shows that insects can be attacked by various methods which can be roughly grouped under three headings. First, very few insects attack plants in full vigour of health, so that any

Agricultural or Forestry methods tending to improve the general health of the crop will be inimical to insect pests. Thus, growing the plants upon the soils best suited to them will tend to make them vigorous and more resistant, so that not continually sowing a crop upon the same ground will tend to defeat the insect in two ways: (1) by maintaining the vigour of the crop, and (2) by starving those insects which remain upon the ground after their food plant has been removed. It is possible also by adding various substances to the soil to increase the vigour of the crop.

Various insects tend to hide under rubbish and may be able to feed upon various wild plants in the absence of the crop and thus, by clean cultivation —clearing up rubbish and keeping the ground as clear as possible of weeds—we can keep down insect attacks.

By a knowledge of the life history of the insect pest it may be possible to grow a crop through its susceptible stage, either before the insect is about or after it has passed, by the choice of early or late varieties or by early or late sowing or by adopting means of hastening the maturing of the crop. Thus, in various places cotton is seriously damaged by a caterpillar known as the Pink Boll-worm, the larva of a minute moth. This caterpillar eats out the contents of the seeds, so that no lint or cotton is developed in the fruit capsule. In Egypt they dodge the attack by withdrawing the irrigation water from the plants after they are well grown and by cutting away the upper shading leaves and branches so as to thoroughly expose the heart of the plant to the sun's rays. This causes the plants to mature more rapidly and too early for the egg-laying of the moth.

A study of the habits of the pest may reveal that, although ready to feed upon a particular crop, it greatly prefers some other plant. In such a case the crop may be largely saved from attack by planting amongst it occasional rows of the favoured plant. Thus a very serious pest of cotton in the United States is a caterpillar known as the Cotton Boll-worm. It is almost cosmopolitan throughout the cultivated areas of the world and attacks various kinds of crops. In certain parts it seems to prefer sweet corn and other varieties of maize so that, by planting a few rows of this at intervals across a cotton field, the cotton becomes almost immune from attack. Such a crop, sown to attract a pest from a more valuable crop, is known as a 'trap crop.'

Observations of the habits of insects sometimes show that they prefer certain varieties of crop to others. Unfortunately, it usually seems to be the poorer varieties which are less attacked, but it may be a question of growing a poor variety or none at all.

The knowledge that insects conceal themselves during the day enables us to set traps for them in the form of bundles of hay or boards laid upon the ground and such kinds of shelters which can be cleared each morning. The knowledge that many insects conceal themselves or their eggs in the cracks in the bark of trees in order to pass the winter leads us to 'wash' the bark with various insect-killing solutions or even to paint the trunks with lime-wash which will dry and thus kill the insects which become embedded in it. The knowledge that various insects feed upon the leafage leads us to spray the leaves with poisons which, harmless to the tree, will destroy the insects which absorb them.

As an example of the value of a detailed knowledge of the habits of insects, I may mention a well-known pest of the apple, the Codlin moth. The little moth, early in the season when the buds are opening, first appears from the pupal stage and lays its eggs, which resemble batches of fish scales, upon the young apple leaves. These hatch while the apple blossom is in bloom and the little caterpillars crawl out upon the flowers and, just as the petals are falling, they eat their way through the nectaries of the flower into the young fruits, inside which they live until they are full-grown. By this time the apple has enlarged and has perhaps fallen in consequence of the damage done by the caterpillar inside it, but, in any case, the full-grown caterpillar leaves the apple and reaches the ground where, under loose bark on the base of the tree or under rubbish lying there, it spins a silken cocoon. In some parts of the world it quickly turns to a pupa and two or three generations may be produced in the year, while in other parts the caterpillar rests in its cocoon during the winter and pupates in the following spring.

Knowing the habits of this insect, it is obvious that it is a difficult pest to attack, but, as it is a very serious one, almost cosmopolitan in its occurrence and in some places damages 90 per cent. of the apple crop, it has been necessary to devise some method of control and this consists in spraying the fruit with poison, just at the time when the petals are falling. It is of course necessary that the poison should reach the place at which the newly-hatched caterpillar will bite its way into the fruit, that is, inside the calyx and, to achieve this, it is better that the spray should fall from above rather than shoot up from below. One other chance presents itself of attacking this pest and that is by gathering and using or destroying the apples which fall during the season so that, in case they contain caterpillars, these shall not have a chance of escaping and changing to pupæ. The clearing up of refuse beneath the trees and keeping the trunks clear of moss and lichens—methods of clean cultivation—also cause the destruction of many pupæ.

We have seen then that Agricultural and Forestry methods and Chemical methods may be adopted for controlling insect pests, but there is another method which depends upon the principle of encouraging internecine strife. This is usually called the 'Biological method' of control and consists in encouraging and aiding other animals, fungi and disease germs to attack the pest.

The living things occupying any particular locality form a community of very intimate relationships. There are, for instance, animals which feed upon the vegetation and there are other animals which feed upon the plant-eating forms. Now if anything were to happen to the vegetation both the groups of animals would be affected. If the flesh-eaters were reduced in numbers the plant-eaters would increase and would tend to reduce and even destroy their food supply, so that we describe the existence of a community of animals and plants as depending upon the 'Balance of Nature.' And this balance of nature is very interesting because things which appear to be in no way related are really dependent upon one another for their existence. For instance, if you are told that the amount of clover seed obtainable from a crop in any district depends upon the number of cats in that district you will have difficulty in understanding the connection, and yet Darwin made it quite clear. Only humble-bees transfer pollen from one clover flower to another so that humble-bees are essential for the pollination of these flowers. The humble-bees nest either just beneath the surface of the ground or in hollows on the surface and their nests are destroyed by field mice which feed upon the stored honey. It has been estimated that more than two-thirds of the nests are thus destroyed all over England. The number of mice is largely dependent upon the number of cats. If the cats increase in numbers the mice will diminish and fewer humble-bee nests will be destroyed and consequently more clover seed will be produced, and thus if the cats diminish in numbers less clover seed will be produced. As it has been remarked that old spinsters are more in the habit of keeping cats than other people, we might extend the biological chain and say that the amount of clover seed produced depends upon the number of old maids.

All living things are thus intimately dependent upon others for their survival and, although changes are always going on in any district, on the whole a balance is maintained, so that the sudden occurrence of any one kind of animal or plant in great abundance is evidence of some interference with the balance of nature having occurred.

That there is, in most cases, a very large death-rate to maintain the balance is obvious when we remember the large numbers of offspring which are produced. For instance, a pike lays about a million eggs and yet the pike is not a particularly abundant fish. Think of the number of caterpillars growing up in the webs of the social kinds and yet these social forms are not crowding out everything else. If we enquire why, for instance, these social caterpillars are unable to win the world for themselves we find that, among other things, they are attacked by a number of parasites which kill off a large proportion of them. These parasites are mainly insects whose larvæ develop inside the caterpillars, feeding upon the living tissues. If anything were to happen to these parasites, we should have an outbreak of social caterpillars, and whenever we get an outbreak of any insect it is due to some interference with the balance of nature.

Now man is always interfering. He drains marshes, thus killing out marsh-loving plants and the animals associated with them. He irrigates dry land, he cuts down forests and by this latter not only interferes with the life of the forest plants and animals but, if he works over a large area, even changes the climate, because forest areas tend to be wetter than others.

Consequently, amongst other effects of his interference, outbreaks of various insects occur and many of these, by reason of the fact that man supplies them with easily obtained food by growing large areas of crops, live largely at his expense. If nature were given time, she would undoubtedly restore the balance and kill off the excessive numbers of individuals but, so long as man continues to cultivate land, nature is not given a fair chance and man, therefore, has hit upon the idea of assisting nature by encouraging and cultivating animals, fungi and disease germs which will help him to destroy those insects which damage his crops.

For instance, many birds feed upon insects and man encourages them by providing nesting boxes and affording the birds protection. A study of the habits of the pest enables him to discover the insect parasites which attack it and these can often be increased in numbers by keeping the pests alive under conditions which will allow the parasites to escape but not the hosts. If we destroy large numbers of pupæ, let us say, of a moth pest, we are at the same time destroying any parasites which those pupæ may contain. If, however, we collect the pupæ and place them in a box covered with gauze through which the parasites can escape when they hatch out but which will prevent the moths from escaping, we are assisting natural control as well as applying artificial control.

This method was actually suggested as long ago as 1871 and, in 1880, an experiment to test it was carried out in France where, in the district of Picardy, great damage was being done in the apple orchards by the Apple-blossom weevil, which lays its egg in the young blossom and whose larva destroys the inside, so that no fruit will set. Evidence of the presence of the larva is given by the fact that the unopened blossom turns a russet-brown colour and is known as a 'capped' blossom. The usual method of control had been to pick off and destroy these capped blossoms and thus to destroy the weevil larvæ and pupæ contained—but, in so doing, no account had been taken of possible parasites attacking the larvæ and thus the parasites were being destroyed as well. The experiment was tried in an isolated area of orchard containing about eight hundred apple trees, all the capped blossoms being picked and placed in boxes covered in with wire-gauze, the mesh of which was too fine to allow any beetles to escape but permitted the free passage of parasites. By this means, it was calculated that in the one year about one million beetles were destroyed and about two hundred and fifty thousand parasites were set free. The process was repeated in the following year and after that, for about ten years, it was found unnecessary to take any steps to deal with this pest, because it was so kept in control by the parasites that it did but little damage.

Since that time and especially within the last few years, this method has been employed in other directions and it is undoubtedly a very important method of control of insect pests.

Now the vast majority of insect pests in any country are those which have been introduced, usually accidentally, from some other country and it is to be noted that, as a rule, an insect is not a pest in its native country—unless it be upon some crop which has been introduced—which is in agreement with what has been said about the balance of nature.

It therefore occurred to certain people that if we could trace the country from which a pest originally came, we could discover the parasites which, in its native country, prevented it from being a pest. We could then introduce those parasites into the pest's new home and we might then expect the pest to be reduced in numbers. The argument is plausible but is not really sound, although a number of extra-ordinarily successful results have been obtained.

The objection to the method lies in the complexity of the relationships of organisms in a community. Although the parasite fits perfectly into its surroundings in its native country and has, as perhaps its main occupation, the keeping down of the numbers of the insect which has become a pest elsewhere, in the new environment the economic conditions of the parasite are necessarily entirely different and it may find dozens of ways of occupying itself without concentrating upon its original natural host.

However, it is not my object here to enter into controversial matters but to emphasize how the study of habits of insects has been the basis of all man's methods of dealing with his insect enemies.

It may be urged that what I have brought forward in this final lecture of the course justifies, to a limited extent, the study of the habits of insects, that is, if the study of the insect can be shown to be useful, then and then only is it justified.

But let me once again urge upon you that this is the wrong view to take of it. The study of insects justifies itself and, however useless it may appear to be, its interest alone will repay the student and that is the spirit of the man of science.

The one objection to it is the fact that there is a struggle for existence and we have to earn our daily bread and, in this commercial world, utility is the only test of money value. It is for this reason that I wish to bring to your notice the fact that there is now an opportunity of earning one's daily bread in the study of insects, in that the recognition of the immense toll taken by them of our lives and our crops has induced Governments and, in some cases, private companies, to employ entomologists to control these pests, so that, in addition to Medical Entomology which has, for a number of years, been a recognised profession, Economic Entomology is now earning a livelihood for a number of enthusiasts.

Although, therefore, I should deplore the suggestion that anyone might take up Entomology with a view to making a living, I shall welcome to the special courses at Cambridge anyone who, having a real interest in insects, decides upon making their study the means by which he shall be able to devote his life to them.

INDEX

For EU product safety concerns, contact us at Calle de José Abascal, 56–1°,
28003 Madrid, Spain or eugpsr@cambridge.org.

www.ingramcontent.com/pod-product-compliance
Ingram Content Group UK Ltd.
Pitfield, Milton Keynes, MK11 3LW, UK
UKHW010850090126
466816UK00011B/141